Feynman Diagrams

" Diagramming The Behavior of Subatomic Particles "

Edited by Paul F. Kisak

Contents

Chapter 1

Feynman diagram

In theoretical physics, **Feynman diagrams** are pictorial representations of the mathematical expressions describing the behavior of subatomic particles. The scheme is named for its inventor, American physicist Richard Feynman, and was first introduced in 1948. The interaction of sub-atomic particles can be complex and difficult to understand intuitively. Feynman diagrams give a simple visualization of what would otherwise be a rather arcane and abstract formula. As David Kaiser writes, "since the middle of the 20th century, theoretical physicists have increasingly turned to this tool to help them undertake critical calculations", and as such "Feynman diagrams have revolutionized nearly every aspect of theoretical physics".[1] While the diagrams are applied primarily to quantum field theory, they can also be used in other fields, such as solid-state theory.

Feynman used Ernst Stueckelberg's interpretation of the positron as if it were an electron moving backward in time.[2] Thus, antiparticles are represented as moving backward along the time axis in Feynman diagrams.

The calculation of probability amplitudes in theoretical particle physics requires the use of rather large and complicated integrals over a large number of variables. These integrals do, however, have a regular structure, and may be represented graphically as Feynman diagrams.

A Feynman diagram is a contribution of a particular class of particle paths, which join and split as described by the diagram. More precisely, and technically, a Feynman diagram is a graphical representation of a perturbative contribution to the transition amplitude or correlation function of a quantum mechanical or statistical field theory. Within the canonical formulation of quantum field theory, a Feynman diagram represents a term in the Wick's expansion of the perturbative S-matrix. Alternatively, the path integral formulation of quantum field theory represents the transition amplitude as a weighted sum of all possible histories of the system from the initial to the final state, in terms of either particles or fields. The transition amplitude is then given as the matrix element of the S-matrix between the initial and the final states of the quantum system.

1.1 Motivation and history

When calculating scattering cross-sections in particle physics, the interaction between particles can be described by starting from a free field that describes the incoming and outgoing particles, and including an interaction Hamiltonian to describe how the particles deflect one another. The amplitude for scattering is the sum of each possible interaction history over all possible intermediate particle states. The number of times the interaction Hamiltonian acts is the order of the perturbation expansion, and the time-dependent perturbation theory for fields is known as the Dyson series. When the intermediate states at intermediate times are energy eigenstates (collections of particles with a definite momentum) the series is called old-fashioned perturbation theory.

The Dyson series can be alternatively rewritten as a sum over Feynman diagrams, where at each interaction vertex both

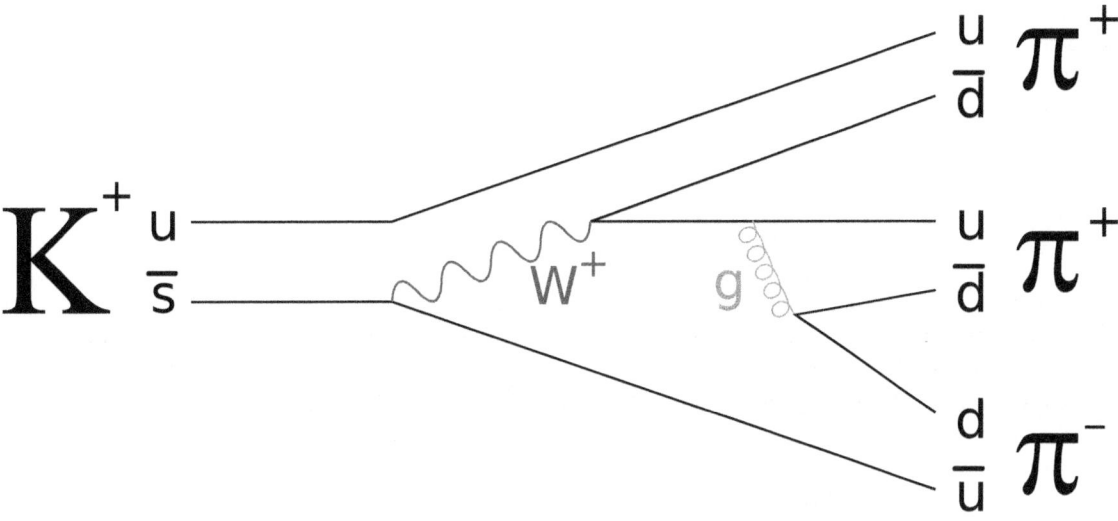

In this diagram, a kaon, made of an up and anti-strange quark, decays both weakly and strongly into three pions, with intermediate steps involving a W boson and a gluon (represented by the blue sine wave and green spiral, respectively).

the energy and momentum are conserved, but where the length of the energy momentum four vector is not equal to the mass. The Feynman diagrams are much easier to keep track of than old-fashioned terms, because the old-fashioned way treats the particle and antiparticle contributions as separate. Each Feynman diagram is the sum of exponentially many old-fashioned terms, because each internal line can separately represent either a particle or an antiparticle. In a non-relativistic theory, there are no antiparticles and there is no doubling, so each Feynman diagram includes only one term.

Feynman gave a prescription for calculating the amplitude for any given diagram from a field theory Lagrangian—the Feynman rules. Each internal line corresponds to a factor of the corresponding virtual particle's propagator; each vertex where lines meet gives a factor derived from an interaction term in the Lagrangian, and incoming and outgoing lines carry an energy, momentum, and spin.

In addition to their value as a mathematical tool, Feynman diagrams provide deep physical insight into the nature of particle interactions. Particles interact in every way available; in fact, intermediate virtual particles are allowed to propagate faster than light. The probability of each final state is then obtained by summing over all such possibilities. This is closely tied to the functional integral formulation of quantum mechanics, also invented by Feynman–see path integral formulation.

The naïve application of such calculations often produces diagrams whose amplitudes are infinite, because the short-distance particle interactions require a careful limiting procedure, to include particle self-interactions. The technique of renormalization, suggested by Ernst Stueckelberg and Hans Bethe and implemented by Dyson, Feynman, Schwinger, and Tomonaga compensates for this effect and eliminates the troublesome infinities. After renormalization, calculations using Feynman diagrams match experimental results with very high accuracy.

Feynman diagram and path integral methods are also used in statistical mechanics and can even be applied to classical mechanics.[3]

1.1.1 Alternative names

Murray Gell-Mann always referred to Feynman diagrams as **Stueckelberg diagrams**, after a Swiss physicist, Ernst Stueckelberg, who devised a similar notation many years earlier. Stueckelberg was motivated by the need for a manifestly covariant formalism for quantum field theory, but did not provide as automated a way to handle symmetry factors and loops, although he was first to find the correct physical interpretation in terms of forward and backward in time particle paths, all without the path-integral.[4] Historically they were sometimes called **Feynman–Dyson diagrams** or **Dyson graphs**,[5] because when they were introduced the path integral was unfamiliar, and Freeman Dyson's derivation from old-fashioned perturbation theory was easier to follow for physicists trained in earlier methods. However, in 2006

Dyson himself stated that the diagrams should be called *Feynman diagrams* because "he taught us how to use them". This reflects historical fact: Feynman had to lobby hard for the diagrams which confused the establishment physicists trained in equations and graphs.[6]

1.2   Representation of physical reality

In their presentations of fundamental interactions,[7][8] written from the particle physics perspective, Gerard 't Hooft and Martinus Veltman gave good arguments for taking the original, non-regularized Feynman diagrams as the most succinct representation of our present knowledge about the physics of quantum scattering of fundamental particles. Their motivations are consistent with the convictions of James Daniel Bjorken and Sidney Drell:[9]

> The Feynman graphs and rules of calculation summarize quantum field theory in a form in close contact with the experimental numbers one wants to understand. Although the statement of the theory in terms of graphs may imply perturbation theory, use of graphical methods in the many-body problem shows that this formalism is flexible enough to deal with phenomena of nonperturbative characters ... Some modification of the Feynman rules of calculation may well outlive the elaborate mathematical structure of local canonical quantum field theory ...

So far there are no opposing opinions. In quantum field theories the Feynman diagrams are obtained from Lagrangian by Feynman rules.

Dimensional regularization is a method for regularizing integrals in the evaluation of Feynman diagrams; it assigns values to them that are meromorphic functions of an auxiliary complex parameter d, called the dimension. Dimensional regularization writes a Feynman integral as an integral depending on the spacetime dimension d and spacetime points.

1.3   Particle-path interpretation

A Feynman diagram is a representation of quantum field theory processes in terms of particle paths. The particle trajectories are represented by the lines of the diagram, which can be squiggly or straight, with an arrow or without, depending on the type of particle. A point where lines connect to other lines is an interaction vertex, and this is where the particles meet and interact: by emitting or absorbing new particles, deflecting one another, or changing type.

There are three different types of lines: *internal lines* connect two vertices, *incoming lines* extend from "the past" to a vertex and represent an initial state, and *outgoing lines* extend from a vertex to "the future" and represent the final state. Sometimes, the bottom of the diagram is the past and the top the future; other times, the past is to the left and the future to the right. When calculating correlation functions instead of scattering amplitudes, there is no past and future and all the lines are internal. The particles then begin and end on little x's, which represent the positions of the operators whose correlation is being calculated.

Feynman diagrams are a pictorial representation of a contribution to the total amplitude for a process that can happen in several different ways. When a group of incoming particles are to scatter off each other, the process can be thought of as one where the particles travel over all possible paths, including paths that go backward in time.

Feynman diagrams are often confused with spacetime diagrams and bubble chamber images because they all describe particle scattering. Feynman diagrams are graphs that represent the trajectories of particles in intermediate stages of a scattering process. Unlike a bubble chamber picture, only the sum of all the Feynman diagrams represent any given particle interaction; particles do not choose a particular diagram each time they interact. The law of summation is in accord with the principle of superposition—every diagram contributes to the total amplitude for the process.

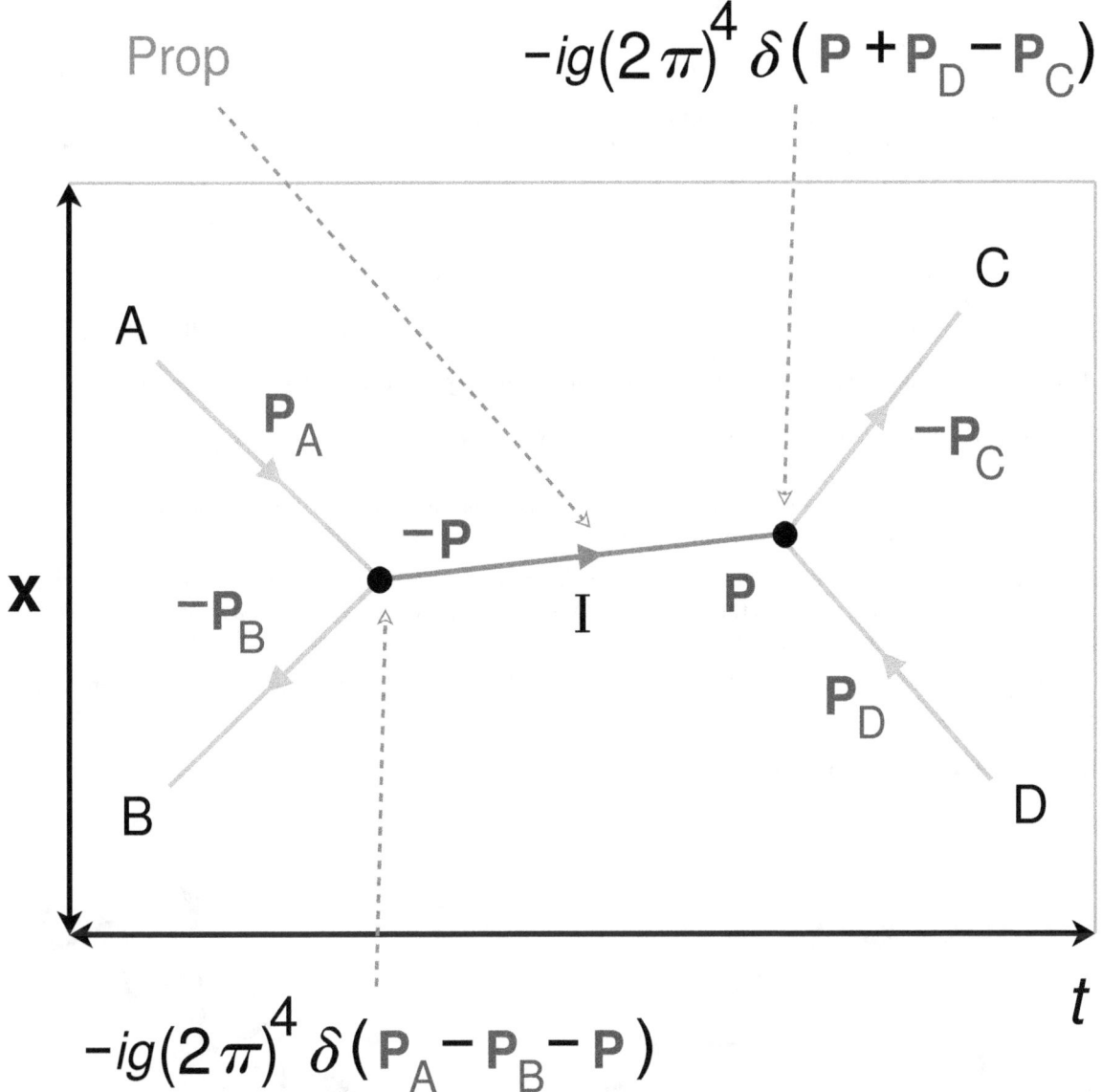

General features of the scattering process $A + B \rightarrow C + D$:
• *internal lines (**red**) for intermediate particles and processes, which has a propagator factor ("prop"), external lines (**orange**) for incoming/outgoing particles to/from vertices (**black**),*
• *at each vertex there is 4-momentum conservation using delta functions, 4-momenta entering the vertex are positive while those leaving are negative, the factors at each vertex and internal line are multiplied in the amplitude integral,*
• *space **x** and time t axes are not always shown, directions of external lines correspond to passage of time.*

1.4   Description

A Feynman diagram represents a perturbative contribution to the amplitude of a quantum transition from some initial quantum state to some final quantum state.

For example, in the process of electron-positron annihilation the initial state is one electron and one positron, the final state: two photons.

The initial state is often assumed to be at the left of the diagram and the final state at the right (although other conventions are also used quite often).

A Feynman diagram consists of points, called vertices, and lines attached to the vertices.

The particles in the initial state are depicted by lines sticking out in the direction of the initial state (e.g., to the left), the particles in the final state are represented by lines sticking out in the direction of the final state (e.g., to the right).

In QED there are two types of particles: electrons/positrons (called fermions) and photons (called gauge bosons). They are represented in Feynman diagrams as follows:

1. Electron in the initial state is represented by a solid line with an arrow pointing toward the vertex ($\rightarrow\bullet$).

2. Electron in the final state is represented by a line with an arrow pointing away from the vertex: ($\bullet\rightarrow$).

3. Positron in the initial state is represented by a solid line with an arrow pointing away from the vertex: ($\leftarrow\bullet$).

4. Positron in the final state is represented by a line with an arrow pointing toward the vertex: ($\bullet\leftarrow$).

5. Photon in the initial and the final state is represented by a wavy line ($\sim\bullet$ and $\bullet\sim$).

In QED a vertex always has three lines attached to it: one bosonic line, one fermionic line with arrow toward the vertex, and one fermionic line with arrow away from the vertex.

The vertices might be connected by a bosonic or fermionic propagator. A bosonic propagator is represented by a wavy line connecting two vertices ($\bullet\sim\bullet$). A fermionic propagator is represented by a solid line (with an arrow in one or another direction) connecting two vertices, ($\bullet\leftarrow\bullet$).

The number of vertices gives the order of the term in the perturbation series expansion of the transition amplitude.

1.4.1   Electron/positron annihilation example

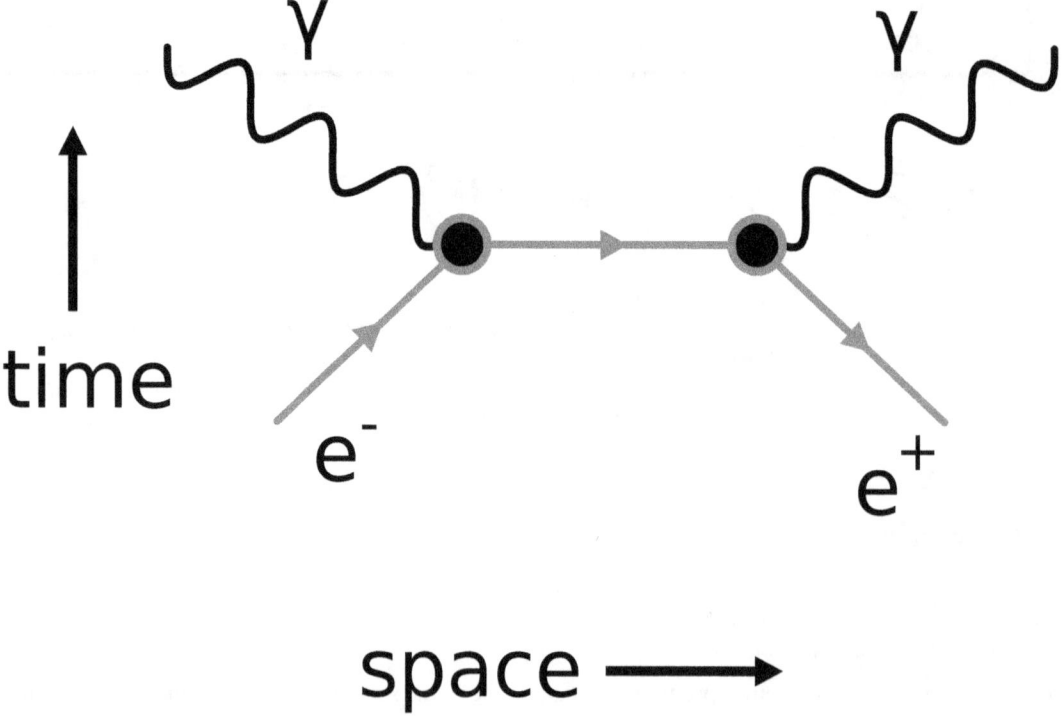

Feynman diagram of electron/positron annihilation

The electron/positron annihilation interaction:

$e^+e^- \rightarrow 2\gamma$

has a contribution from the second order Feynman diagram shown adjacent:

In the initial state (at the bottom; early time) there is one electron (e^-) and one positron (e^+) and in the final state (at the top; late time) there are two photons (γ).

1.5   Canonical quantization formulation

The probability amplitude for a transition of a quantum system from the initial state $|i\rangle$ to the final state $|f\rangle$ is given by the matrix element

$$S_{fi} = \langle f|S|i\rangle \,,$$

where S is the S-matrix.

In the canonical quantum field theory the S-matrix is represented within the interaction picture by the perturbation series in the powers of the interaction Lagrangian,

$$S = \sum_{n=0}^{\infty} \frac{i^n}{n!} \int \prod_{j=1}^{n} d^4x_j T \prod_{j=1}^{n} L_v(x_j) \equiv \sum_{n=0}^{\infty} S^{(n)} \,,$$

where L_v is the interaction Lagrangian and T signifies the time-ordered product of operators.

A Feynman diagram is a graphical representation of a term in the Wick's expansion of the time-ordered product in the n-th order term $S^{(n)}$ of the S-matrix,

$$T \prod_{j=1}^{n} L_v(x_j) = \sum_{\text{all possible contractions}} (\pm)N \prod_{j=1}^{n} L_v(x_j) \,,$$

where N signifies the normal-product of the operators and ($\pm$) takes care of the possible sign change when commuting the fermionic operators to bring them together for a contraction (a propagator).

1.5.1   Feynman rules

The diagrams are drawn according to the Feynman rules, which depend upon the interaction Lagrangian. For the QED interaction Lagrangian, $L_v = -g\bar{\psi}\gamma^\mu\psi A_\mu$, describing the interaction of a fermionic field ψ with a bosonic gauge field A_μ, the Feynman rules can be formulated in coordinate space as follows:

1. Each integration coordinate x_j is represented by a point (sometimes called a vertex);

2. A bosonic propagator is represented by a wiggly line connecting two points;

3. A fermionic propagator is represented by a solid line connecting two points;

4. A bosonic field $A_\mu(x_i)$ is represented by a wiggly line attached to the point x_i ;

5. A fermionic field $\psi(x_i)$ is represented by a solid line attached to the point x_i with an arrow toward the point;

6. A fermionic field $\bar{\psi}(x_i)$ is represented by a solid line attached to the point x_i with an arrow from the point;

1.5.2 Example: second order processes in QED

The second order perturbation term in the S-matrix is

$$S^{(2)} = \frac{(ie)^2}{2!} \int d^4x \, d^4x' \, T\bar{\psi}(x) \, \gamma^\mu \, \psi(x) \, A_\mu(x) \, \bar{\psi}(x') \, \gamma^\nu \, \psi(x') \, A_\nu(x').$$

Scattering of fermions

The Wick's expansion of the integrand gives (among others) the following term

$$N\bar{\psi}(x)\gamma^\mu\psi(x)\bar{\psi}(x')\gamma^\nu\psi(x')\underline{A_\mu(x)A_\nu(x')} \ ,$$

where

$$\underline{A_\mu(x)A_\nu(x')} = \int \frac{d^4k}{(2\pi)^4} \frac{-ig_{\mu\nu}}{k^2+i0} e^{-ik(x-x')}$$

is the electromagnetic contraction (propagator) in the Feynman gauge. This term is represented by the Feynman diagram at the right. This diagram gives contributions to the following processes:

1. e^-e^- scattering (initial state at the right, final state at the left of the diagram);

2. e^+e^+ scattering (initial state at the left, final state at the right of the diagram);

3. e^-e^+ scattering (initial state at the bottom/top, final state at the top/bottom of the diagram).

Compton scattering and annihilation/generation of e^-e^+ pairs

Another interesting term in the expansion is

$$N\bar{\psi}(x) \, \gamma^\mu \, \underline{\psi(x) \, \bar{\psi}(x')} \, \gamma^\nu \, \psi(x') \, A_\mu(x) \, A_\nu(x') \ ,$$

where

$$\underline{\psi(x)\bar{\psi}(x')} = \int \frac{d^4p}{(2\pi)^4} \frac{i}{\gamma p - m + i0} e^{-ip(x-x')}$$

is the fermionic contraction (propagator).

1.6 Path integral formulation

In a path-integral, the field Lagrangian, integrated over all possible field histories, defines the probability amplitude to go from one field configuration to another. In order to make sense, the field theory should have a well-defined ground state, and the integral should be performed a little bit rotated into imaginary time, i.e. a Wick Rotation.

1.6.1 Scalar field Lagrangian

A simple example is the free relativistic scalar field in d-dimensions, whose action integral is:

$$S = \int \frac{1}{2} \partial_\mu \phi \partial^\mu \phi d^d x \ .$$

The probability amplitude for a process is:

$$\int_A^B e^{iS} D\phi \,,$$

where A and B are space-like hypersurfaces that define the boundary conditions. The collection of all the $\phi(A)$ on the starting hypersurface give the initial value of the field, analogous to the starting position for a point particle, and the field values $\phi(B)$ at each point of the final hypersurface defines the final field value, which is allowed to vary, giving a different amplitude to end up at different values. This is the field-to-field transition amplitude.

The path integral gives the expectation value of operators between the initial and final state:

$$\int_A^B e^{iS} \phi(x_1)...\phi(x_n) D\phi = \langle A|\phi(x_1)...\phi(x_n)|B\rangle \,,$$

and in the limit that A and B recede to the infinite past and the infinite future, the only contribution that matters is from the ground state (this is only rigorously true if the path-integral is defined slightly rotated into imaginary time). The path integral should be thought of as analogous to a probability distribution, and it is convenient to define it so that multiplying by a constant doesn't change anything:

$$\frac{\int e^{iS} \phi(x_1)...\phi(x_n) D\phi}{\int e^{iS} D\phi} = \langle 0|\phi(x_1)....\phi(x_n)|0\rangle \,.$$

The normalization factor on the bottom is called the *partition function* for the field, and it coincides with the statistical mechanical partition function at zero temperature when rotated into imaginary time.

The initial-to-final amplitudes are ill-defined if one thinks of the continuum limit right from the beginning, because the fluctuations in the field can become unbounded. So the path-integral should be thought of as on a discrete square lattice, with lattice spacing a and the limit $a \to 0$ should be taken carefully. If the final results do not depend on the shape of the lattice or the value of a, then the continuum limit exists.

1.6.2   On a lattice

On a lattice, (i), the field can be expanded in Fourier modes:

$$\phi(x) = \int \frac{dk}{(2\pi)^d} \phi(k) e^{ik \cdot x} = \int_k \phi(k) e^{ikx} \,.$$

Here the integration domain is over k restricted to a cube of side length $2\pi/a$, so that large values of k are not allowed. It is important to note that the k-measure contains the factors of 2π from Fourier transforms, this is the best standard convention for k-integrals in QFT. The lattice means that fluctuations at large k are not allowed to contribute right away, they only start to contribute in the limit $a \to 0$. Sometimes, instead of a lattice, the field modes are just cut off at high values of k instead.

It is also convenient from time to time to consider the space-time volume to be finite, so that the k modes are also a lattice. This is not strictly as necessary as the space-lattice limit, because interactions in k are not localized, but it is convenient for keeping track of the factors in front of the k-integrals and the momentum-conserving delta functions that will arise.

On a lattice, (ii), the action needs to be discretized:

$$S = \sum_{<x,y>} \frac{1}{2}(\phi(x) - \phi(y))^2 \,,$$

where $< x, y >$ is a pair of nearest lattice neighbors x and y. The discretization should be thought of as defining what the derivative $\partial_\mu \phi$ means.

In terms of the lattice Fourier modes, the action can be written:

$$S = \int_k ((1 - \cos(k_1)) + (1 - \cos(k_2)) + ... + (1 - \cos(k_d))) \phi_k^* \phi^k \,.$$

For k near zero this is:

$$S = \int_k \frac{1}{2} k^2 |\phi(k)|^2 \,.$$

Now we have the continuum Fourier transform of the original action. In finite volume, the quantity $d^d k$ is not infinitesimal, but becomes the volume of a box made by neighboring Fourier modes, or $(2\pi/V)^d$.

The field ϕ is real-valued, so the Fourier transform obeys:

$$\phi(k)^* = \phi(-k) \,.$$

In terms of real and imaginary parts, the real part of $\phi(k)$ is an even function of k, while the imaginary part is odd. The Fourier transform avoids double-counting, so that it can be written:

$$S = \int_k \frac{1}{2} k^2 \phi(k) \phi(-k)$$

over an integration domain that integrates over each pair (k,−k) exactly once.

For a complex scalar field with action

$$S = \int \frac{1}{2} \partial_\mu \phi^* \partial^\mu \phi \, d^d x$$

the Fourier transform is unconstrained:

$$S = \int_k \frac{1}{2} k^2 |\phi(k)|^2$$

and the integral is over all k.

Integrating over all different values of $\phi(x)$ is equivalent to integrating over all Fourier modes, because taking a Fourier transform is a unitary linear transformation of field coordinates. When you change coordinates in a multidimensional integral by a linear transformation, the value of the new integral is given by the determinant of the transformation matrix. If

$$y_i = A_{ij}x_j \, ,$$

then

$$\det(A) \int dx_1 dx_2 ... dx_n = \int dy_1 dy_2 ... dy_n \, .$$

If A is a rotation, then

$$A^T A = I$$

so that $\det A = \pm 1$, and the sign depends on whether the rotation includes a reflection or not.

The matrix that changes coordinates from $\phi(x)$ to $\phi(k)$ can be read off from the definition of a Fourier transform.

$$A_{kx} = e^{ikx}$$

and the Fourier inversion theorem tells you the inverse:

$$A_{kx}^{-1} = e^{-ikx}$$

which is the complex conjugate-transpose, up to factors of 2π . On a finite volume lattice, the determinant is nonzero and independent of the field values.

$$\det A = 1$$

and the path integral is a separate factor at each value of k.

$$\int \exp\left(\frac{i}{2} \sum_k k^2 \phi^*(k)\phi(k) \right) D\phi = \prod_k \int_{\phi_k} e^{\frac{i}{2}k^2|\phi_k|^2 d^d k}$$

The factor $d^d k$ is the infinitesimal volume of a discrete cell in k-space, in a square lattice box $d^d k = 1/L^d$, where L is the side-length of the box. Each separate factor is an oscillatory Gaussian, and the width of the Gaussian diverges as the volume goes to infinity.

In imaginary time, the *Euclidean action* becomes positive definite, and can be interpreted as a probability distribution. The probability of a field having values ϕ_k is

$$e^{\int_k -\frac{1}{2}k^2 \phi_k^* \phi_k} = \prod_k e^{-k^2|\phi_k|^2 d^d k}$$

The expectation value of the field is the statistical expectation value of the field when chosen according to the probability distribution:

$$\langle \phi(x_1)...\phi(x_n) \rangle = \frac{\int e^{-S} \phi(x_1)...\phi(x_n) D\phi}{\int e^{-S} D\phi}$$

Since the probability of ϕ_k is a product, the value of $\phi(k)$ at each separate value of k is independently Gaussian distributed. The variance of the Gaussian is 1/ (k^2 d^dk), which is formally infinite, but that just means that the fluctuations are unbounded in infinite volume. In any finite volume, the integral is replaced by a discrete sum, and the variance of the integral is V/k^2 .

1.6.3   Monte Carlo

The path integral defines a probabilistic algorithm to generate a Euclidean scalar field configuration. Randomly pick the real and imaginary parts of each Fourier mode at wavenumber k to be a gaussian random variable with variance $1/k^2$. This generates a configuration $\phi_C(k)$ at random, and the Fourier transform gives $\phi_C(x)$. For real scalar fields, the algorithm must generate only one of each pair $\phi(k), \phi(-k)$, and make the second the complex conjugate of the first.

To find any correlation function, generate a field again and again by this procedure, and find the statistical average:

$$\langle \phi(x_1)...\phi(x_n) \rangle = \lim_{|C| \to \infty} \frac{\sum_C \phi_C(x_1)...\phi_C(x_n)}{|C|}$$

where $|C|$ is the number of configurations, and the sum is of the product of the field values on each configuration. The Euclidean correlation function is just the same as the correlation function in statistics or statistical mechanics. The quantum mechanical correlation functions are an analytic continuation of the Euclidean correlation functions.

For free fields with a quadratic action, the probability distribution is a high-dimensional Gaussian, and the statistical average is given by an explicit formula. But the Monte Carlo method also works well for bosonic interacting field theories where there is no closed form for the correlation functions.

1.6.4   Scalar propagator

Each mode is independently Gaussian distributed. The expectation of field modes is easy to calculate:

$$\langle \phi_k \phi_{k'} \rangle = 0$$

for $k \neq k'$, since then the two Gaussian random variables are independent and both have zero mean.

$$\langle \phi_k \phi_k \rangle = \frac{V}{k^2}$$

in finite volume V, when the two k-values coincide, since this is the variance of the Gaussian. In the infinite volume limit,

$$\langle \phi(k)\phi(k') \rangle = \delta(k - k') \frac{1}{k^2}$$

Strictly speaking, this is an approximation: the lattice propagator is:

$$\langle \phi(k)\phi(k') \rangle = \delta(k - k') \frac{1}{2(d - \cos(k_1) + \cos(k_2)... + \cos(k_d))}$$

But near k=0, for field fluctuations long compared to the lattice spacing, the two forms coincide.

It is important to emphasize that the delta functions contain factors of 2π, so that they cancel out the 2π factors in the measure for k integrals.

$$\delta(k) = (2\pi)^d \delta_D(k_1)\delta_D(k_2)...\delta_D(k_d)$$

where $\delta_D(k)$ is the ordinary one-dimensional Dirac delta function. This convention for delta-functions is not universal—some authors keep the factors of 2π in the delta functions (and in the k-integration) explicit.

1.6.5   Equation of motion

The form of the propagator can be more easily found by using the equation of motion for the field. From the Lagrangian, the equation of motion is:

$$\partial_\mu \partial^\mu \phi = 0$$

and in an expectation value, this says:

$$\partial_\mu \partial^\mu \langle \phi(x)\phi(y) \rangle = 0$$

Where the derivatives act on x, and the identity is true everywhere except when x and y coincide, and the operator order matters. The form of the singularity can be understood from the canonical commutation relations to be a delta-function. Defining the (euclidean) *Feynman propagator* Δ as the Fourier transform of the time-ordered two-point function (the one that comes from the path-integral):

$$\partial^2 \Delta(x) = i\delta(x)$$

So that:

$$\Delta(k) = \frac{i}{k^2}$$

If the equations of motion are linear, the propagator will always be the reciprocal of the quadratic-form matrix that defines the free Lagrangian, since this gives the equations of motion. This is also easy to see directly from the Path integral. The factor of i disappears in the Euclidean theory.

Wick theorem

Main article: Wick's Theorem

Because each field mode is an independent Gaussian, the expectation values for the product of many field modes obeys *Wick's theorem*:

$$\langle \phi(k_1)\phi(k_2)...\phi(k_n) \rangle$$

is zero unless the field modes coincide in pairs. This means that it is zero for an odd number of ϕ 's, and for an even number of ϕ 's, it is equal to a contribution from each pair separately, with a delta function.

$$\langle \phi(k_1)...\phi(k_{2n}) \rangle = \sum \prod_{i,j} \frac{\delta(k_i - k_j)}{k_i^2}$$

where the sum is over each partition of the field modes into pairs, and the product is over the pairs. For example,

$$\langle \phi(k_1)\phi(k_2)\phi(k_3)\phi(k_4) \rangle = \frac{\delta(k_1 - k_2)}{k_1^2}\frac{\delta(k_3 - k_4)}{k_3^2} + \frac{\delta(k_1 - k_3)}{k_3^2}\frac{\delta(k_2 - k_4)}{k_2^2} + \frac{\delta(k_1 - k_4)}{k_1^2}\frac{\delta(k_2 - k_3)}{k_2^2}$$

An interpretation of Wick's theorem is that each field insertion can be thought of as a dangling line, and the expectation value is calculated by linking up the lines in pairs, putting a delta function factor that ensures that the momentum of each partner in the pair is equal, and dividing by the propagator.

Higher Gaussian moments—completing Wick's theorem

There is a subtle point left before Wick's theorem is proved—what if more than two of the phis have the same momentum? If it's an odd number, the integral is zero; negative values cancel with the positive values. But if the number is even, the integral is positive. The previous demonstration assumed that the phis would only match up in pairs.

But the theorem is correct even when arbitrarily many of the phis are equal, and this is a notable property of Gaussian integration:

$$I = \int e^{-ax^2/2} dx = \sqrt{\frac{2\pi}{a}}$$

$$\frac{\partial^n}{\partial a^n} I = \int \frac{x^{2n}}{2^n} e^{-ax^2/2} dx = \frac{1 \cdot 3 \cdot 5... \cdot (2n-1)}{2 \cdot 2 \cdot 2... \quad \cdot 2} \sqrt{2\pi} a^{-\frac{2n+1}{2}}$$

Dividing by I,

$$\langle x^{2n} \rangle = \frac{\int x^{2n} e^{-ax^2/2}}{\int e^{-ax^2/2}} = 1 \cdot 3 \cdot 5... \cdot (2n-1)\frac{1}{a^n}$$

$$\langle x^2 \rangle = \frac{1}{a}$$

If Wick's theorem were correct, the higher moments would be given by all possible pairings of a list of 2n x's:

$$\langle x_1 x_2 x_3 ... x_{2n} \rangle$$

where the x's are all the same variable, the index is just to keep track of the number of ways to pair them. The first x can be paired with 2n−1 others, leaving 2n−2. The next unpaired x can be paired with 2n-3 different x's leaving 2n−4, and so on. This means that Wick's theorem, uncorrected, says that the expectation value of x^{2n} should be:

$$\langle x^{2n} \rangle = (2n - 1) \cdot (2n - 3) .... \cdot 5 \cdot 3 \cdot 1 (\langle x^2 \rangle)^n$$

and this is in fact the correct answer. So Wick's theorem holds no matter how many of the momenta of the internal variables coincide.

Interaction

Interactions are represented by higher order contributions, since quadratic contributions are always Gaussian. The simplest interaction is the quartic self-interaction, with an action:

$$S = \int \partial^\mu \phi \partial_\mu \phi + \frac{\lambda}{4!} \phi^4.$$

The reason for the combinatorial factor 4! will be clear soon. Writing the action in terms of the lattice (or continuum) Fourier modes:

$$S = \int_k k^2 |\phi(k)|^2 + \int_{k_1 k_2 k_3 k_4} \phi(k_1)\phi(k_2)\phi(k_3)\phi(k_4)\delta(k_1 + k_2 + k_3 + k_4) = S_F + X.$$

Where S_F is the free action, whose correlation functions are given by Wick's theorem. The exponential of S in the path integral can be expanded in powers of λ, giving a series of corrections to the free action.

$$e^{-S} = e^{-S_F}(1 + X + \frac{1}{2!}XX + \frac{1}{3!}XXX + ...)$$

The path integral for the interacting action is then a power series of corrections to the free action. The term represented by X should be thought of as four half-lines, one for each factor of $\phi(k)$. The half-lines meet at a vertex, which contributes a delta-function that ensures that the sum of the momenta are all equal.

To compute a correlation function in the interacting theory, there is a contribution from the X terms now. For example, the path-integral for the four-field correlator:

$$\langle \phi(k_1)\phi(k_2)\phi(k_3)\phi(k_4) \rangle = \frac{\int e^{-S}\phi(k_1)\phi(k_2)\phi(k_3)\phi(k_4)D\phi}{Z}$$

which in the free field was only nonzero when the momenta k were equal in pairs, is now nonzero for all values of the k. The momenta of the insertions $\phi(k_i)$ can now match up with the momenta of the X's in the expansion. The insertions should also be thought of as half-lines, four in this case, which carry a momentum k, but one that is not integrated.

The lowest order contribution comes from the first nontrivial term $e^{-S_F} X$ in the Taylor expansion of the action. Wick's theorem requires that the momenta in the X half-lines, the $\phi(k)$ factors in X, should match up with the momenta of the external half-lines in pairs. The new contribution is equal to:

$$\lambda \frac{1}{k_1^2} \frac{1}{k_2^2} \frac{1}{k_3^2} \frac{1}{k_4^2}.$$

The 4! inside X is canceled because there are exactly 4! ways to match the half-lines in X to the external half-lines. Each of these different ways of matching the half-lines together in pairs contributes exactly once, regardless of the values of the k's, by Wick's theorem.

Feynman diagrams

The expansion of the action in powers of X gives a series of terms with progressively higher number of X's. The contribution from the term with exactly n X's are called n-th order.

The n-th order terms has:

1. 4n internal half-lines, which are the factors of $\phi(k)$ from the X's. These all end on a vertex, and are integrated over all possible k.

2. external half-lines, which are the come from the $\phi(k)$ insertions in the integral.

By Wick's theorem, each pair of half-lines must be paired together to make a *line*, and this line gives a factor of

$$\frac{\delta(k_1 + k_2)}{k_1^2}$$

which multiplies the contribution. This means that the two half-lines that make a line are forced to have equal and opposite momentum. The line itself should be labelled by an arrow, drawn parallel to the line, and labeled by the momentum in the line k. The half-line at the tail end of the arrow carries momentum k, while the half-line at the head-end carries momentum −k. If one of the two half-lines is external, this kills the integral over the internal k, since it forces the internal k to be equal to the external k. If both are internal, the integral over k remains.

The diagrams that are formed by linking the half-lines in the X's with the external half-lines, representing insertions, are the Feynman diagrams of this theory. Each line carries a factor of $\frac{1}{k^2}$, the propagator, and either goes from vertex to vertex, or ends at an insertion. If it is internal, it is integrated over. At each vertex, the total incoming k is equal to the total outgoing k.

The number of ways of making a diagram by joining half-lines into lines almost completely cancels the factorial factors coming from the Taylor series of the exponential and the 4! at each vertex.

Loop order

A forest diagram is one where all the internal lines have momentum that is completely determined by the external lines and the condition that the incoming and outgoing momentum are equal at each vertex. The contribution of these diagrams is a product of propagators, without any integration. A tree diagram is a connected forest diagram.

An example of a tree diagram is the one where each of four external lines end on an X. Another is when three external lines end on an X, and the remaining half-line joins up with another X, and the remaining half-lines of this X run off to external lines. These are all also forest diagrams (as every tree is a forest); an example of a forest that is not a tree is when eight external lines end on two X's.

It is easy to verify that in all these cases, the momenta on all the internal lines is determined by the external momenta and the condition of momentum conservation in each vertex.

A diagram that is not a forest diagram is called a *loop* diagram, and an example is one where two lines of an X are joined to external lines, while the remaining two lines are joined to each other. The two lines joined to each other can have any momentum at all, since they both enter and leave the same vertex. A more complicated example is one where two X's are joined to each other by matching the legs one to the other. This diagram has no external lines at all.

The reason loop diagrams are called loop diagrams is because the number of k-integrals that are left undetermined by momentum conservation is equal to the number of independent closed loops in the diagram, where independent loops are counted as in homology theory. The homology is real-valued (actually R^d valued), the value associated with each line is the momentum. The boundary operator takes each line to the sum of the end-vertices with a positive sign at the head and a negative sign at the tail. The condition that the momentum is conserved is exactly the condition that the boundary of the k-valued weighted graph is zero.

A set of k-values can be relabeled whenever there is a closed loop going from vertex to vertex, never revisiting the same vertex. Such a cycle can be thought of as the boundary of a 2-cell. The k-labelings of a graph that conserves momentum (which has zero boundary) up to redefinitions of k (up to boundaries of 2-cells) define the first homology of a graph. The number of independent momenta that are not determined is then equal to the number of independent homology loops. For many graphs, this is equal to the number of loops as counted in the most intuitive way.

Symmetry factors

The number of ways to form a given Feynman diagram by joining together half-lines is large, and by Wick's theorem, each way of pairing up the half-lines contributes equally. Often, this completely cancels the factorials in the denominator of each term, but the cancellation is sometimes incomplete.

The uncancelled denominator is called the *symmetry factor* of the diagram. The contribution of each diagram to the correlation function must be divided by its symmetry factor.

For example, consider the Feynman diagram formed from two external lines joined to one X, and the remaining two half-lines in the X joined to each other. There are 4×3 ways to join the external half-lines to the X, and then there is only one way to join the two remaining lines to each other. The X comes divided by 4!=4×3×2, but the number of ways to link up the X half lines to make the diagram is only 4×3, so the contribution of this diagram is divided by two.

For another example, consider the diagram formed by joining all the half-lines of one X to all the half-lines of another X. This diagram is called a *vacuum bubble*, because it does not link up to any external lines. There are 4! ways to form this diagram, but the denominator includes a 2! (from the expansion of the exponential, there are two X's) and two factors of 4!. The contribution is multiplied by 4!/(2×4!×4!) = 1/48.

Another example is the Feynman diagram formed from two X's where each X links up to two external lines, and the remaining two half-lines of each X are joined to each other. The number of ways to link an X to two external lines is 4×3, and either X could link up to either pair, giving an additional factor of 2. The remaining two half-lines in the two X's can be linked to each other in two ways, so that the total number of ways to form the diagram is 4×3×4×3×2×2, while the denominator is 4!×4!×2!. The total symmetry factor is 2, and the contribution of this diagram is divided by 2.

The symmetry factor theorem gives the symmetry factor for a general diagram: the contribution of each Feynman diagram must be divided by the order of its group of automorphisms, the number of symmetries that it has.

An automorphism of a Feynman graph is a permutation M of the lines and a permutation N of the vertices with the following properties:

1. If a line l goes from vertex v to vertex v', then M(l) goes from N(v) to N(v'). If the line is undirected, as it is for a real scalar field, then M(l) can go from N(v') to N(v) too.

2. If a line l ends on an external line, M(l) ends on the same external line.

3. If there are different types of lines, M(l) should preserve the type.

This theorem has an interpretation in terms of particle-paths: when identical particles are present, the integral over all intermediate particles must not double-count states that differ only by interchanging identical particles.

Proof: To prove this theorem, label all the internal and external lines of a diagram with a unique name. Then form the diagram by linking the a half-line to a name and then to the other half line.

Now count the number of ways to form the named diagram. Each permutation of the X's gives a different pattern of linking names to half-lines, and this is a factor of n!. Each permutation of the half-lines in a single X gives a factor of 4!. So a named diagram can be formed in exactly as many ways as the denominator of the Feynman expansion.

But the number of unnamed diagrams is smaller than the number of named diagram by the order of the automorphism group of the graph.

Connected diagrams: *linked-cluster theorem*

Roughly speaking, a Feynman diagram is called *connected* if all vertices and propagator lines are linked by a sequence of vertices and propagators of the diagram itself. If one views it as a (undirected) graph it is connected. The remarkable relevance of such diagrams in QFTs is due to the fact that they are sufficient to determine the quantum partition function $Z[J]$. More precisely, connected Feynman diagrams determine

$$iW[J] \equiv \ln Z[J].$$

To see this, one should recall that

$$Z[J] \propto \sum_k D_k$$

with D_k constructed from some (arbitrary) Feynman diagram that can be thought to consist of several connected components C_i. If one encounters n_i (identical) copies of a component C_i within the Feynman diagram D_k one has to include a *symmetry factor* $n_i!$. However, in the end each contribution of a Feynman diagram D_k to the partition function has the generic form

$$\prod_i \frac{C_i^{n_i}}{n_i!}$$

where i labels the (infinite) many connected Feynman diagrams possible.

A scheme to successively create such contributions from the D_k to $Z[J]$ is obtained by

$$\left(\frac{1}{0!} + \frac{C_1}{1!} + \frac{C_1^2}{2!} + \dots \right) \left(1 + C_2 + \frac{1}{2}C_2^2 + \dots \right) \dots$$

and therefore yields

$$Z[J] \propto \prod_i \sum_{n_i=0}^{\infty} \frac{C_i^{n_i}}{n_i!} = \exp \sum_i C_i \propto \exp W[J].$$

To establish the *normalization* $Z_0 = \exp W[0] = 1$ one simply calculates all connected *vacuum diagrams*, i.e., the diagrams without any *sources* J (sometimes referred to as *external legs* of a Feynman diagram).

Vacuum bubbles

An immediate consequence of the linked-cluster theorem is that all vacuum bubbles, diagrams without external lines, cancel when calculating correlation functions. A correlation function is given by a ratio of path-integrals:

$$\langle \phi_1(x_1)...\phi_n(x_n) \rangle = \frac{\int e^{-S} \phi_1(x_1)...\phi_n(x_n) D\phi}{\int e^{-S} D\phi}.$$

The top is the sum over all Feynman diagrams, including disconnected diagrams that do not link up to external lines at all. In terms of the connected diagrams, the numerator includes the same contributions of vacuum bubbles as the denominator:

$$\int e^{-S} \phi_1(x_1)...\phi_n(x_n) D\phi = \left(\sum E_i\right)\left(\exp\left(\sum_i C_i\right)\right).$$

Where the sum over E diagrams includes only those diagrams each of whose connected components end on at least one external line. The vacuum bubbles are the same whatever the external lines, and give an overall multiplicative factor. The denominator is the sum over all vacuum bubbles, and dividing gets rid of the second factor.

The vacuum bubbles then are only useful for determining Z itself, which from the definition of the path integral is equal to:

$$Z = \int e^{-S} D\phi = e^{-HT} = e^{-\rho V}$$

where ρ is the energy density in the vacuum. Each vacuum bubble contains a factor of $\delta(k)$ zeroing the total k at each vertex, and when there are no external lines, this contains a factor of $\delta(0)$, because the momentum conservation is over-enforced. In finite volume, this factor can be identified as the total volume of space time. Dividing by the volume, the remaining integral for the vacuum bubble has an interpretation: it is a contribution to the energy density of the vacuum.

Sources

Correlation functions are the sum of the connected Feynman diagrams, but the formalism treats the connected and disconnected diagrams differently. Internal lines end on vertices, while external lines go off to insertions. Introducing *sources* unifies the formalism, by making new vertices where one line can end.

Sources are external fields, fields that contribute to the action, but are not dynamical variables. A scalar field source is another scalar field h that contributes a term to the (Lorentz) Lagrangian:

$$\int h(x)\phi(x)d^dx = \int h(k)\phi(k)d^dk$$

In the Feynman expansion, this contributes H terms with one half-line ending on a vertex. Lines in a Feynman diagram can now end either on an X vertex, or on an H-vertex, and only one line enters an H vertex. The Feynman rule for an H-vertex is that a line from an H with momentum k gets a factor of h(k).

The sum of the connected diagrams in the presence of sources includes a term for each connected diagram in the absence of sources, except now the diagrams can end on the source. Traditionally, a source is represented by a little "x" with one line extending out, exactly as an insertion.

$$\log(Z[h]) = \sum_{n,C} h(k_1)h(k_2)...h(k_n)C(k_1,...,k_n)$$

where $C(k_1,....,k_n)$ is the connected diagram with n external lines carrying momentum as indicated. The sum is over all connected diagrams, as before.

The field h is not dynamical, which means that there is no path integral over h: h is just a parameter in the Lagrangian, which varies from point to point. The path integral for the field is:

$$Z[h] = \int e^{iS+i\int h\phi} D\phi$$

and it is a function of the values of h at every point. One way to interpret this expression is that it is taking the Fourier transform in field space. If there is a probability density on R^n, the Fourier transform of the probability density is:

$$\int \rho(y)e^{iky}d^n y = \langle e^{iky} \rangle = \langle \prod_{i=1}^{n} e^{ih_i y_i} \rangle$$

The Fourier transform is the expectation of an oscillatory exponential. The path integral in the presence of a source h(x) is:

$$Z[h] = \int e^{iS} e^{i\int_x h(x)\phi(x)} D\phi = \langle e^{ih\phi} \rangle$$

which, on a lattice, is the product of an oscillatory exponential for each field value:

$$\langle \prod_x e^{ih_x \phi_x} \rangle$$

The fourier transform of a delta-function is a constant, which gives a formal expression for a delta function:

$$\delta(x-y) = \int e^{ik(x-y)} dk$$

This tells you what a field delta function looks like in a path-integral. For two scalar fields ϕ and η ,

$$\delta(\phi - \eta) = \int e^{ih(x)(\phi(x) - \eta(x)d^dx} Dh$$

Which integrates over the Fourier transform coordinate, over h. This expression is useful for formally changing field coordinates in the path integral, much as a delta function is used to change coordinates in an ordinary multi-dimensional integral.

The partition function is now a function of the field h, and the physical partition function is the value when h is the zero function:

The correlation functions are derivatives of the path integral with respect to the source:

$$\langle\phi(x)\rangle = \frac{1}{Z}\frac{\partial}{\partial h(x)}Z[h] = \frac{\partial}{\partial h(x)}\log(Z[h]).$$

In Euclidean space, source contributions to the action can still appear with a factor of "i", so that they still do a Fourier transform.

1.6.6   Spin 1/2; "photons" and "ghosts"

Spin 1/2: Grassmann integrals

The field path-integral can be extended to the Fermi case, but only if the notion of integration is expanded. A Grassmann integral of a free Fermi field is a high-dimensional determinant or Pfaffian, which defines the new type of Gaussian integration appropriate for Fermi fields.

The two fundamental formulas of Grassmann integration are:

$$\int e^{M_{ij}\bar\psi^i\psi^j} D\bar\psi D\psi = \mathrm{Det}(M)$$

where M is an arbitrary matrix and $\psi, \bar\psi$ are independent Grassmann variables for each index i, and

$$\int e^{\frac{1}{2}A_{ij}\psi^i\psi^j} D\psi = \mathrm{Pfaff}(A)$$

Where A is an antisymmetric matrix, ψ is a collection of Grassmann variables, and the 1/2 is to prevent double-counting (since $\psi^i\psi^j = -\psi^j\psi^i$). In matrix notation, where $\bar\psi$ and $\bar\eta$ are Grassmann valued row vectors, η and ψ are Grassmann valued column vectors, and M is a real valued matrix:

$$Z = \int e^{\bar\psi M\psi + \bar\eta\psi + \bar\psi\eta} D\bar\psi D\psi = \int e^{(\bar\psi + \bar\eta M^{-1})M(\psi + M^{-1}\eta) - \bar\eta M^{-1}\eta} D\bar\psi D\psi = \mathrm{Det}(M)e^{-\bar\eta M^{-1}\eta}$$

Where the last equality is a consequence of the translation invariance of the Grassmann integral. The Grassmann variables η are external sources for ψ, and differentiating with respect to η pulls down factors of $\bar\psi$.

$$\langle \bar{\psi}\psi \rangle = \frac{1}{Z}\frac{\partial}{\partial \eta}\frac{\partial}{\partial \bar{\eta}}Z|_{\eta=\bar{\eta}=0} = M^{-1}$$

again, in a schematic matrix notation. The meaning of the formula above is that the derivative with respect to the appropriate component of η and $\bar{\eta}$ gives the matrix element of M^{-1}. This is exactly analogous to the Bosonic path integration formula for a Gaussian integral of a complex Bosonic field:

$$\int e^{\phi^* M\phi + h^*\phi + \phi^* h} D\phi^* D\phi = \frac{e^{h^* M^{-1}h}}{\mathrm{Det}(M)}$$

$$\langle \phi^*\phi \rangle = \frac{1}{Z}\frac{\partial}{\partial h}\frac{\partial}{\partial h^*}Z|_{h=h^*=0} = M^{-1}$$

So that the propagator is the inverse of the matrix in the quadratic part of the action in both the Bose and Fermi case.

For real Grassmann fields, for Majorana fermions, the path integral a Pfaffian times a source quadratic form, and the formulas give the square root of the determinant, just as they do for real Bosonic fields. The propagator is still the inverse of the quadratic part.

The free Dirac Lagrangian:

$$\int \bar{\psi}(\gamma^\mu \partial_\mu - m)\psi$$

formally gives the equations of motion and the anticommutation relations of the Dirac field, just as the Klein Gordon Lagrangian in an ordinary path integral gives the equations of motion and commutation relations of the scalar field. By using the spatial Fourier transform of the Dirac field as a new basis for the Grassmann algebra, the quadratic part of the Dirac action becomes simple to invert:

$$S = \int_k \bar{\psi}(i\gamma^\mu k_\mu - m)\psi.$$

The propagator is the inverse of the matrix M linking $\psi(k)$ and $\bar{\psi}(k)$, since different values of k do not mix together.

$$\langle \bar{\psi}(k')\psi(k) \rangle = \delta(k+k')\frac{1}{\gamma \cdot k - m} = \delta(k+k')\frac{\gamma \cdot k + m}{k^2 - m^2}$$

The analog of Wick's theorem matches psi and psi-bars in pairs:

$$\langle \bar{\psi}(k_1)\bar{\psi}(k_2)...\bar{\psi}(k_n)\psi(k_1')...\psi(k_n) \rangle = \sum_{\text{pairings}} (-1)^S \prod_{\text{pairs } i,j} \delta(k_i - k_j)\frac{1}{\gamma \cdot k_i - m}$$

where S is the sign of the permutation that reorders the sequence of psi-bars and psis to put the ones that are paired up to make the delta-functions next to each other, with the psi-bar coming right before the psi. Since a psi-psi-bar pair is a commuting element of the Grassmann algebra, it doesn't matter what order the pairs are in. If more than one psi/psi-bar pair have the same k, the integral is zero, and it is easy to check that the sum over pairings gives zero in this case (there are always an even number of them). This is the Grassmann analog of the higher Gaussian moments that completed the Bosonic Wick's theorem earlier.

The rules for spin-1/2 Dirac particles are as follows: The propagator is the inverse of the Dirac operator, the lines have arrows just as for a complex scalar field, and the diagram acquires an overall factor of −1 for each closed Fermi loop. If there are an odd number of Fermi loops, the diagram changes sign. Historically, the −1 rule was very difficult for Feynman to discover. He discovered it after a long process of trial and error, since he lacked a proper theory of Grassmann integration.

The rule follows from the observation that the number of Fermi lines at a vertex is always even. Each term in the Lagrangian must always be Bosonic. A Fermi loop is counted by following Fermionic lines until one comes back to the starting point, then removing those lines from the diagram. Repeating this process eventually erases all the Fermionic lines: this is the Euler algorithm to 2-color a graph, which works whenever each vertex has even degree. Note that the number of steps in the Euler algorithm is only equal to the number of independent Fermionic homology cycles in the common special case that all terms in the Lagrangian are exactly quadratic in the Fermi fields, so that each vertex has exactly two Fermionic lines. When there are four-Fermi interactions (like in the Fermi effective theory of the Weak interactions) there are more k-integrals than Fermi loops. In this case, the counting rule should apply the Euler algorithm by pairing up the Fermi lines at each vertex into pairs that together form a bosonic factor of the term in the Lagrangian, and when entering a vertex by one line, the algorithm should always leave with the partner line.

To clarify and prove the rule, consider a Feynman diagram formed from vertices, terms in the Lagrangian, with Fermion fields. The full term is Bosonic, it is a commuting element of the Grassmann algebra, so the order in which the vertices appear is not important. The Fermi lines are linked into loops, and when traversing the loop, one can reorder the vertex terms one after the other as one goes around without any sign cost. The exception is when you return to the starting point, and the final half-line must be joined with the unlinked first half-line. This requires one permutation to move the last psi-bar to go in front of the first psi, and this gives the sign.

This rule is the only visible effect of the exclusion principle in internal lines. When there are external lines, the amplitudes are antisymmetric when two Fermi insertions for identical particles are interchanged. This is automatic in the source formalism, because the sources for Fermi fields are themselves Grassmann valued.

Spin 1: photons

The naive propagator for photons is infinite, since the Lagrangian for the A-field is:

$$S = \int \frac{1}{4} F^{\mu\nu} F_{\mu\nu} = \int -\frac{1}{2} (\partial^\mu A_\nu \partial_\mu A^\nu - \partial^\mu A_\mu \partial_\nu A^\nu).$$

The quadratic form defining the propagator is non-invertible. The reason is the gauge invariance of the field, adding a gradient to A does not change the physics.

To fix this problem, one needs to fix a gauge. The most convenient way is to demand that the divergence of A is some function f, whose value is random from point to point. It does no harm to integrate over the values of f, since it only determines the choice of gauge. This procedure inserts the following factor into the path integral for A:

$$\int \delta(\partial_\mu A^\mu - f) e^{-\frac{f^2}{2}} Df.$$

The first factor, the delta function, fixes the gauge. The second factor sums over different values of f that are inequivalent gauge fixings. This is simply

$$e^{-\frac{(\partial_\mu A_\mu)^2}{2}}.$$

The additional contribution from gauge-fixing cancels the second half of the free Lagrangian, giving the Feynman Lagrangian:

$$S = \int \partial^\mu A^\nu \partial_\mu A_\nu$$

which is just like four independent free scalar fields, one for each component of A. The Feynman propagator is:

$$\langle A_\mu(k) A_\nu(k') \rangle = \delta(k + k') \frac{g_{\mu\nu}}{k^2}.$$

The one difference is that the sign of one propagator is wrong in the Lorentz case: the timelike component has an opposite sign propagator. This means that these particle states have negative norm—they are not physical states. In the case of photons, it is easy to show by diagram methods that these states are not physical—their contribution cancels with longitudinal photons to only leave two physical photon polarization contributions for any value of k.

If the averaging over f is done with a coefficient different from 1/2, the two terms don't cancel completely. This gives a covariant Lagrangian with a coefficient λ, which does not affect anything:

$$S = \int \frac{1}{2} (\partial^\mu A^\nu \partial_\mu A_\nu - \lambda (\partial_\mu A^\mu)^2)$$

and the covariant propagator for QED is:

$$\langle A_\mu(k) A_\nu(k') \rangle = \delta(k + k') \frac{g_{\mu\nu} - \lambda \frac{k_\mu k_\nu}{k^2}}{k^2}.$$

Spin 1: nonabelian ghosts

To find the Feynman rules for nonabelian Gauge fields, the procedure that performs the Gauge fixing must be carefully corrected to account for a change of variables in the path-integral.

The gauge fixing factor has an extra determinant from popping the delta function:

$$\delta(\partial_\mu A_\mu - f) e^{-\frac{f^2}{2}} \mathrm{Det} M$$

To find the form of the determinant, consider first a simple two-dimensional integral of a function f that depends only on r, not on the angle θ. Inserting an integral over theta:

$$\int f(r) dx dy = \int f(r) \int d\theta \delta(y) |\frac{dy}{d\theta}| dx dy$$

The derivative-factor ensures that popping the delta function in θ removes the integral. Exchanging the order of integration,

$$\int f(r)dxdy = \int d\theta \int f(r)\delta(y)|\frac{dy}{d\theta}|dxdy$$

but now the delta-function can be popped in y,

$$\int f(r)dxdy = \int d\theta_0 \int f(x)|\frac{dy}{d\theta}|dx\,.$$

The integral over θ just gives an overall factor of 2π , while the rate of change of y with a change in θ is just x, so this exercise reproduces the standard formula for polar integration of a radial function:

$$\int f(r)dxdy = 2\pi \int f(x)xdx$$

In the path-integral for a nonabelian gauge field, the analogous manipulation is:

$$\int DA \int \delta(F(A))\text{Det}(\frac{\partial F}{\partial G})DGe^{iS} = \int DG \int \delta(F(A))\text{Det}(\frac{\partial F}{\partial G})e^{iS}$$

The factor in front is the volume of the gauge group, and it contributes a constant, which can be discarded. The remaining integral is over the gauge fixed action.

$$\int \text{Det}(\frac{\partial F}{\partial G})e^{iS_{GF}}DA$$

To get a covariant gauge, the gauge fixing condition is the same as in the Abelian case:

$$\partial_\mu A^\mu = f$$

Whose variation under an infinitesimal gauge transformation is given by:

$$\partial_\mu D_\mu \alpha$$

where α is the adjoint valued element of the Lie algebra at every point that performs the infinitesimal gauge transformation. This adds the Faddeev Popov determinant to the action:

$$Det(\partial_\mu D_\mu)$$

which can be rewritten as a Grassman integral by introducing ghost fields:

$$\int e^{\bar{\eta}\partial_\mu D^\mu \eta} D\bar{\eta}D\eta$$

The determinant is independent of f, so the path-integral over f can give the Feynman propagator (or a covariant propagator) by choosing the measure for f as in the abelian case. The full gauge fixed action is then the Yang Mills action in Feynman gauge with an additional ghost action:

$$S = \int Tr\partial_\mu A_\nu \partial^\mu A^\nu + f^i_{jk}\partial^\nu A^\mu_i A^j_\mu A^k_\nu + f^i_{jr}f^r_{kl}A_iA_jA^kA^l + Tr\partial_\mu\bar{\eta}\partial^\mu\eta + \bar{\eta}A_j\eta$$

The diagrams are derived from this action. The propagator for the spin-1 fields has the usual Feynman form. There are vertices of degree 3 with momentum factors whose couplings are the structure constants, and vertices of degree 4 whose couplings are products of structure constants. There are additional ghost loops, which cancel out timelike and longitudinal states in A loops.

In the Abelian case, the determinant for covariant gauges does not depend on A, so the ghosts do not contribute to the connected diagrams.

1.7 Particle-path representation

Feynman diagrams were originally discovered by Feynman, by trial and error, as a way to represent the contribution to the S-matrix from different classes of particle trajectories.

1.7.1 Schwinger representation

The Euclidean scalar propagator has a suggestive representation:

$$\frac{1}{p^2+m^2} = \int_0^\infty e^{-\tau(p^2+m^2)}d\tau$$

The meaning of this identity (which is an elementary integration) is made clearer by Fourier transforming to real space.

$$\Delta(x) = \int_0^\infty d\tau e^{-m^2\tau}\frac{1}{(4\pi\tau)^{d/2}}e^{\frac{-x^2}{4\tau}}$$

The contribution at any one value of τ to the propagator is a Gaussian of width $\sqrt{\tau}$. The total propagation function from 0 to x is a weighted sum over all proper times τ of a normalized Gaussian, the probability of ending up at x after a random walk of time τ.

The path-integral representation for the propagator is then:

$$\Delta(x) = \int_0^\infty d\tau \int DX e^{-\int_0^\tau (\dot{x}^2/2+m^2)d\tau'}$$

which is a path-integral rewrite of the Schwinger representation.

The Schwinger representation is both useful for making manifest the particle aspect of the propagator, and for symmetrizing denominators of loop diagrams.

1.7.2   Combining denominators

The Schwinger representation has an immediate practical application to loop diagrams. For example, For the diagram in the phi-4 theory formed by joining two x's together in two half-lines, and making the remaining lines external, the integral over the internal propagators in the loop is:

$$\int_k \frac{1}{(k^2 + m^2)} \frac{1}{((k+p)^2 + m^2)} \ .$$

Here one line carries momentum k and the other k+p. The asymmetry can be fixed by putting everything in the Schwinger representation.

$$\int_{t,t'} e^{-t(k^2+m^2)-t'((k+p)^2+m^2)} dt dt' \ .$$

Now the exponent mostly depends on t+t',

$$\int_{t,t'} e^{-(t+t')(k^2+m^2)-t'2p\cdot k-t'p^2} \ ,$$

except for the asymmetrical little bit. Defining the variable u=(t+t') and $v = t'/u$, the variable u goes from 0 to infinity, while v goes from 0 to 1. The variable u is the total proper time for the loop, while v parametrizes the fraction of the proper time on the top of the loop vs. the bottom.

The Jacobian for this transformation of variables is easy to work out from the identities:

$$d(uv) = dt' \quad du = dt + dt' \ ,$$

and "wedging" gives

$$u du \wedge dv = dt \wedge dt'$$

This allows the u integral to be evaluated explicitly:

$$\int_{u,v} u e^{-u(k^2+m^2+v2p\cdot k+vp^2)} = \int \frac{1}{(k^2 + m^2 + v2p\cdot k - vp^2)^2} dv$$

leaving only the v -integral. This method, invented by Schwinger but usually attributed to Feynman, is called *combining denominator*. Abstractly, it is the elementary identity:

$$\frac{1}{AB} = \int_0^1 \frac{1}{(vA + (1-v)B)^2} dv$$

But this form does not provide the physical motivation for introducing v — v is the proportion of proper time on one of the legs of the loop.

Once the denominators are combined, a shift in k to $k' = k + vp$ symmetrizes everything:

$$\int_0^1 \int \frac{1}{(k^2 + m^2 + v2p \cdot k + vp^2)^2} dk dv = \int_0^1 \int \frac{1}{(k'^2 + m^2 + v(1-v)p^2)^2} dk' dv$$

This form shows that the moment that p^2 is more negative than 4 times the mass of the particle in the loop, which happens in a physical region of Lorentz space, the integral has a cut. This is exactly when the external momentum can create physical particles.

When the loop has more vertices, there are more denominators to combine:

$$\int dk \frac{1}{(k^2 + m^2)} \frac{1}{((k+p_1)^2 + m^2)} \cdots \frac{1}{((k+p_n)^2 + m^2)}$$

The general rule follows from the Schwinger prescription for n+1 denominators:

$$\frac{1}{D_0 D_1 ... D_n} = \int_0^\infty \cdots \int_0^\infty e^{-u_0 D_0 ... - u_n D_n} du_0 ... du_n .$$

The integral over the Schwinger parameters u_i can be split up as before into an integral over the total proper time $u = u_0 + u_1 ... + u_n$ and an integral over the fraction of the proper time in all but the first segment of the loop $v_i = u_i/u$ for $i \in \{1, 2, ..., n\}$. The v's are positive and add up to less than 1, so that the v integral is over an n dimensional simplex.

The Jacobian for the coordinate transformation can be worked out as before:

$$du = du_0 + du_1 ... + du_n$$
$$d(uv_i) = du_i .$$

"Wedging" all these equation together, one obtains

$$u^n du \wedge dv_1 \wedge dv_2 ... \wedge dv_n = du_0 \wedge du_1 ... \wedge du_n .$$

This gives the integral:

$$\int_0^\infty \int_{\text{simplex}} u^n e^{-u(v_0 D_0 + v_1 D_1 + v_2 D_2 ... + v_n D_n)} dv_1 ... dv_n du ,$$

where the simplex is the region defined by the conditions $v_i > 0$ and $\sum_{i=1}^n v_i < 1$ as well as $v_0 = 1 - \sum_{i=1}^n v_i$. Performing the u integral gives the general prescription for combining denominators:

$$\frac{1}{D_0 ... D_n} = n! \int_{\text{simplex}} \frac{1}{(v_0 D_0 + v_1 D_1 ... + v_n D_n)^{n+1}} dv_1 dv_2 ... dv_n$$

Since the numerator of the integrand is not involved, the same prescription works for any loop, no matter what the spins are carried by the legs. The interpretation of the parameters v_i is that they are the fraction of the total proper time spent on each leg.

1.7.3 Scattering

The correlation functions of a quantum field theory describe the scattering of particles. The definition of "particle" in relativistic field theory is not self-evident, because if you try to determine the position so that the uncertainty is less than the compton wavelength, the uncertainty in energy is large enough to produce more particles and antiparticles of the same type from the vacuum. This means that the notion of a single-particle state is to some extent incompatible with the notion of an object localized in space.

In the 1930s, Wigner gave a mathematical definition for single-particle states: they are a collection of states that form an irreducible representation of the Poincaré group. Single particle states describe an object with a finite mass, a well defined momentum, and a spin. This definition is fine for protons and neutrons, electrons and photons, but it excludes quarks, which are permanently confined, so the modern point of view is more accommodating: a particle is anything whose interaction can be described in terms of Feynman diagrams, which have an interpretation as a sum over particle trajectories.

A field operator can act to produce a one-particle state from the vacuum, which means that the field operator $\phi(x)$ produces a superposition of Wigner particle states. In the free field theory, the field produces one particle states only. But when there are interactions, the field operator can also produce 3-particle, 5-particle (if there is no +/− symmetry also 2, 4, 6 particle) states too. To compute the scattering amplitude for single particle states only requires a careful limit, sending the fields to infinity and integrating over space to get rid of the higher-order corrections.

The relation between scattering and correlation functions is the LSZ-theorem: The scattering amplitude for n particles to go to m-particles in a scattering event is the given by the sum of the Feynman diagrams that go into the correlation function for n+m field insertions, leaving out the propagators for the external legs.

For example, for the $\lambda\phi^4$ interaction of the previous section, the order λ contribution to the (Lorentz) correlation function is:

$$\langle\phi(k_1)\phi(k_2)\phi(k_3)\phi(k_4)\rangle = \frac{i}{k_1^2}\frac{i}{k_2^2}\frac{i}{k_3^2}\frac{i}{k_4^2}i\lambda$$

Stripping off the external propagators, that is, removing the factors of i/k^2 , gives the invariant scattering amplitude M:

$$M = i\lambda$$

which is a constant, independent of the incoming and outgoing momentum. The interpretation of the scattering amplitude is that the sum of $|M|^2$ over all possible final states is the probability for the scattering event. The normalization of the single-particle states must be chosen carefully, however, to ensure that M is a relativistic invariant.

Non-relativistic single particle states are labeled by the momentum k, and they are chosen to have the same norm at every value of k. This is because the nonrelativistic unit operator on single particle states is:

$$\int dk|k\rangle\langle k|$$

In relativity, the integral over the k-states for a particle of mass m integrates over a hyperbola in E,k space defined by the energy–momentum relation:

$$E^2 - k^2 = m^2$$

If the integral weighs each k point equally, the measure is not Lorentz invariant. The invariant measure integrates over all values of k and E, restricting to the hyperbola with a Lorentz invariant delta function:

$$\int \delta(E^2 - k^2 - m^2)|E, k\rangle\langle E, k|dEdk = \int \frac{dk}{2E}|k\rangle\langle k|$$

So the normalized k-states are different from the relativistically normalized k-states by a factor of $\sqrt{E} = (k^2 - m^2)^{\frac{1}{4}}$

The invariant amplitude M is then the probability amplitude for relativistically normalized incoming states to become relativistically normalized outgoing states.

For nonrelativistic values of k, the relativistic normalization is the same as the nonrelativistic normalization (up to a constant factor $\sqrt{m}$). In this limit, the ϕ^4 invariant scattering amplitude is still constant. The particles created by the field phi scatter in all directions with equal amplitude.

The nonrelativistic potential, which scatters in all directions with an equal amplitude (in the Born approximation), is one whose Fourier transform is constant—a delta-function potential. The lowest order scattering of the theory reveals the non-relativistic interpretation of this theory—it describes a collection of particles with a delta-function repulsion. Two such particles have an aversion to occupying the same point at the same time.

1.8 Nonperturbative effects

Thinking of Feynman diagrams as a perturbation series, nonperturbative effects like tunneling do not show up, because any effect that goes to zero faster than any polynomial does not affect the Taylor series. Even bound states are absent, since at any finite order particles are only exchanged a finite number of times, and to make a bound state, the binding force must last forever.

But this point of view is misleading, because the diagrams not only describe scattering, but they also are a representation of the short-distance field theory correlations. They encode not only asymptotic processes like particle scattering, they also describe the multiplication rules for fields, the operator product expansion. Nonperturbative tunneling processes involve field configurations that on average get big when the coupling constant gets small, but each configuration is a coherent superposition of particles whose local interactions are described by Feynman diagrams. When the coupling is small, these become collective processes that involve large numbers of particles, but where the interactions between each of the particles is simple.

This means that nonperturbative effects show up asymptotically in resummations of infinite classes of diagrams, and these diagrams can be locally simple. The graphs determine the local equations of motion, while the allowed large-scale configurations describe non-perturbative physics. But because Feynman propagators are nonlocal in time, translating a field process to a coherent particle language is not completely intuitive, and has only been explicitly worked out in certain special cases. In the case of nonrelativistic bound states, the Bethe–Salpeter equation describes the class of diagrams to include to describe a relativistic atom. For quantum chromodynamics, the Shifman Vainshtein Zakharov sum rules describe non-perturbatively excited long-wavelength field modes in particle language, but only in a phenomenological way.

The number of Feynman diagrams at high orders of perturbation theory is very large, because there are as many diagrams as there are graphs with a given number of nodes. Nonperturbative effects leave a signature on the way in which the number of diagrams and resummations diverge at high order. It is only because non-perturbative effects appear in hidden form in diagrams that it was possible to analyze nonperturbative effects in string theory, where in many cases a Feynman description is the only one available.

1.9 In popular culture

- The use of the above diagram of the virtual particle producing a quark–antiquark pair was featured in the television sit-com *The Big Bang Theory*, in the episode *"The Bat Jar Conjecture"*.

- *PhD Comics* of January 11, 2012, shows Feynman diagrams that *visualize and describe quantum academic interactions*, i.e. the paths followed by Ph.D. students when interacting with their advisors[10]

1.10   See also

- Julian Schwinger#Schwinger and Feynman

- Stueckelberg–Feynman interpretation

- Invariance mechanics

- Penguin diagram

- Path integral formulation

- Propagator

- List of Feynman diagrams

- Angular momentum diagrams (quantum mechanics)

1.11   Notes

[1] "Physics and Feynman's Diagrams" by David Kaiser, *American Scientist*, Volume 93, p. 156

[2] Feynman, Richard (1949). "The Theory of Positrons".*Physical Review***76**(76): 749.Bibcode:1949PhRv...76..749F.doi:10.11
In this solution, the "negative energy states" appear in a form which may be pictured (as by Stückelberg) in space-time as waves traveling away from the external potential backwards in time. Experimentally, such a wave corresponds to a positron approaching the potential and annihilating the electron.

[3] R. Penco, D. Mauro (2006). "Perturbation theory via Feynman diagrams in classical mechanics". arXiv:hep-th/0605061v2.

[4] George Johnson (July 2000). "The Jaguar and the Fox". *The Atlantic*. Retrieved February 26, 2013.

[5] Gribbin, John and Mary. *Richard Feynman: A Life in Science*, Penguin-Putnam, 1997 Ch 5.

[6] Leonard Mlodinow. *Feynman's Rainbow*. Vintage, 2011. p. 29

[7] Gerardus 't Hooft, Martinus Veltman, *Diagrammar*, CERN Yellow Report 1973, reprinted in G. 't Hooft, *Under the Spell of Gauge Principle* (World Scientific, Singapore, 1994), Introduction online

[8] Martinus Veltman, *Diagrammatica: The Path to Feynman Diagrams*, Cambridge Lecture Notes in Physics, ISBN 0-521-45692-4

[9] Bjorken, J. D.; Drell, S. D. (1965). "Relativistic Quantum Fields". New York: McGraw-Hill: viii.

[10] Jorge Cham, Academic Interaction - Feynman Diagrams, January 11, 2012.

1.12   References

- Gerardus 't Hooft, Martinus Veltman, *Diagrammar*, CERN Yellow Report 1973, online

- David Kaiser, *Drawing Theories Apart: The Dispersion of Feynman Diagrams in Postwar Physics*, Chicago: University of Chicago Press, 2005. ISBN 0-226-42266-6

- Martinus Veltman, *Diagrammatica: The Path to Feynman Diagrams*, Cambridge Lecture Notes in Physics, ISBN 0-521-45692-4 (expanded, updated version of above)

- Mark Srednicki, *Quantum Field Theory*, online Script (2006)

1.13 External links

- AMS article: "What's New in Mathematics: Finite-dimensional Feynman Diagrams"

- Draw Feynman diagrams explained by Flip Tanedo at Quantumdiaries.com

- Drawing Feynman diagrams with FeynDiagram C++ library that produces PostScript output.

- Feynman Diagram Examples using Thorsten Ohl's Feynmf LaTeX package.

- Online Diagram Tool A graphical application for creating publication ready diagrams.

- JaxoDraw A Java program for drawing Feynman diagrams.

- SCaViS – a Java program that can be used for drawing Feynman diagrams using Python scripts

- Bowley, Roger; Copeland, Ed (2010). "Feynman Diagrams". *Sixty Symbols*. Brady Haran for the University of Nottingham.

Chapter 2

Antiparticle

Corresponding to most kinds of particles, there is an associated antimatter **antiparticle** with the same mass and opposite charge (including electric charge). For example, the antiparticle of the electron is the positively charged positron, which is produced naturally in certain types of radioactive decay.

The laws of nature are very nearly symmetrical with respect to particles and antiparticles. For example, an antiproton and a positron can form an antihydrogen atom, which is believed to have the same properties as a hydrogen atom. This leads to the question of why the formation of matter after the Big Bang resulted in a universe consisting almost entirely of matter, rather than being a half-and-half mixture of matter and antimatter. The discovery of Charge Parity violation helped to shed light on this problem by showing that this symmetry, originally thought to be perfect, was only approximate.

Particle-antiparticle pairs can annihilate each other, producing photons; since the charges of the particle and antiparticle are opposite, total charge is conserved. For example, the positrons produced in natural radioactive decay quickly annihilate themselves with electrons, producing pairs of gamma rays, a process exploited in positron emission tomography.

Antiparticles are produced naturally in beta decay, and in the interaction of cosmic rays in the Earth's atmosphere.

Because charge is conserved, it is not possible to create an antiparticle without either destroying a particle of the same charge (as in beta decay, when a proton (positive charge) is destroyed, a neutron created and a positron (positive charge, antiparticle) is also created and emitted) or by creating a particle of the opposite charge. The latter is seen in many processes in which both a particle and its antiparticle are created simultaneously, as in particle accelerators. This is the inverse of the particle-antiparticle annihilation process.

Although particles and their antiparticles have opposite charges, electrically neutral particles need not be identical to their antiparticles. The neutron, for example, is made out of quarks, the antineutron from antiquarks, and they are distinguishable from one another because neutrons and antineutrons annihilate each other upon contact. However, other neutral particles are their own antiparticles, such as photons, hypothetical gravitons, and some WIMPs.

2.1 History

2.1.1 Experiment

In 1932, soon after the prediction of positrons by Paul Dirac, Carl D. Anderson found that cosmic-ray collisions produced these particles in a cloud chamber— a particle detector in which moving electrons (or positrons) leave behind trails as they move through the gas. The electric charge-to-mass ratio of a particle can be measured by observing the radius of curling of its cloud-chamber track in a magnetic field. Positrons, because of the direction that their paths curled, were at first mistaken for electrons travelling in the opposite direction. Positron paths in a cloud-chamber trace the same helical path as an electron but rotate in the opposite direction with respect to the magnetic field direction due to their having the same magnitude of charge-to-mass ratio but with opposite charge and, therefore, opposite signed charge-to-mass ratios.

The antiproton and antineutron were found by Emilio Segrè and Owen Chamberlain in 1955 at the University of Califor-

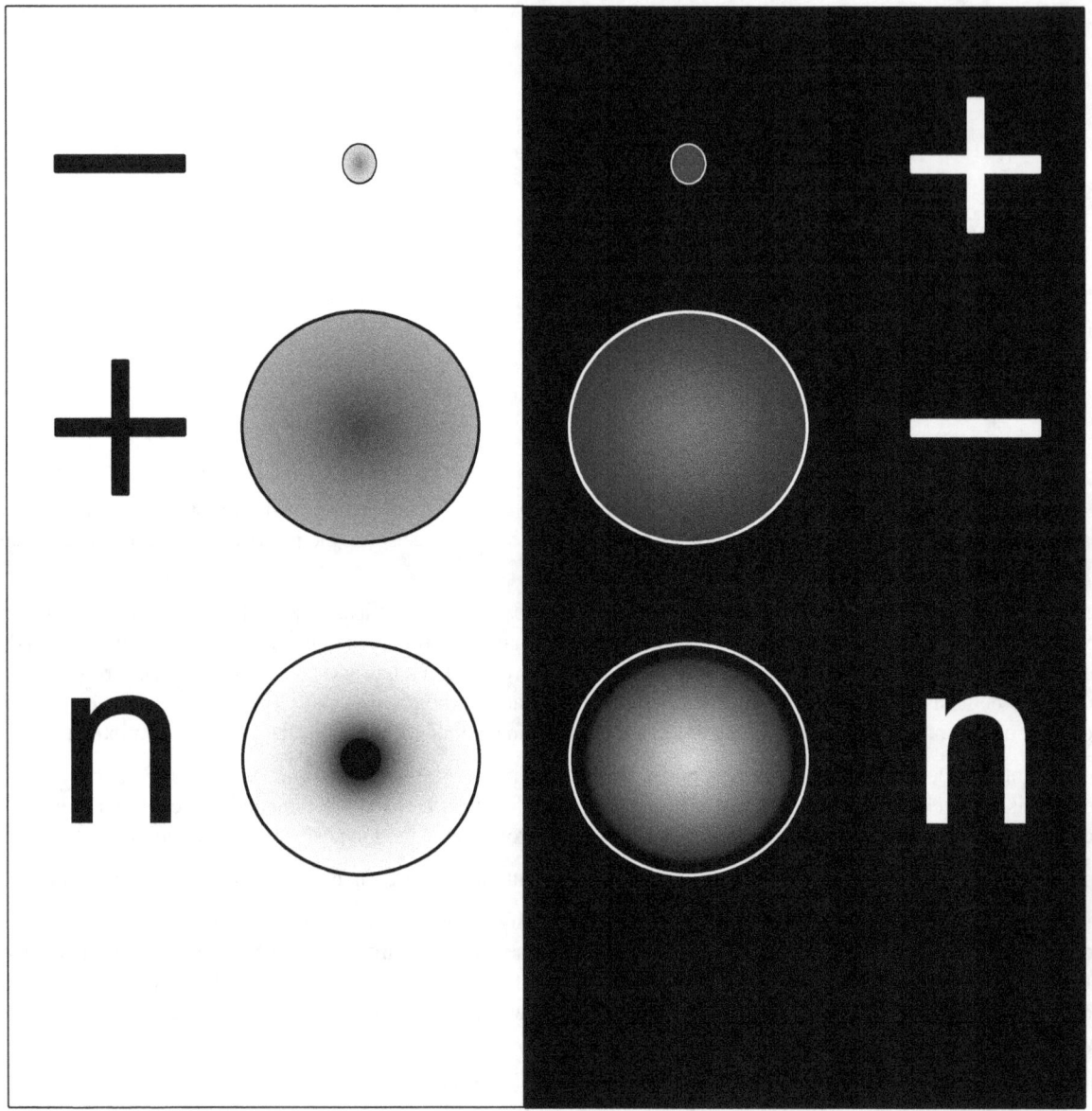

Illustration of electric charge of particles (left) and antiparticles (right). From top to bottom; electron/positron, proton/antiproton, neutron/antineutron.

nia, Berkeley. Since then, the antiparticles of many other subatomic particles have been created in particle accelerator experiments. In recent years, complete atoms of antimatter have been assembled out of antiprotons and positrons, collected in electromagnetic traps.[1]

2.1.2 Dirac Hole theory

... the development of quantum field theory made the interpretation of antiparticles as holes unnecessary, even though it lingers on in many textbooks.

Steven Weinberg[2]

Solutions of the Dirac equation contained negative energy quantum states. As a result, an electron could always radiate energy and fall into a negative energy state. Even worse, it could keep radiating infinite amounts of energy because there were infinitely many negative energy states available. To prevent this unphysical situation from happening, Dirac proposed that a "sea" of negative-energy electrons fills the universe, already occupying all of the lower-energy states so that, due to the Pauli exclusion principle, no other electron could fall into them. Sometimes, however, one of these negative-energy particles could be lifted out of this Dirac sea to become a positive-energy particle. But, when lifted out, it would leave behind a *hole* in the sea that would act exactly like a positive-energy electron with a reversed charge. These he interpreted as "negative-energy electrons" and attempted to identify them with protons in his 1930 paper *A Theory of Electrons and Protons*[3] However, these "negative-energy electrons" turned out to be positrons, and not protons.

This picture implied an infinite negative charge for the universe—a problem of which Dirac was aware. Dirac tried to argue that we would perceive this as the normal state of zero charge. Another difficulty was the difference in masses of the electron and the proton. Dirac tried to argue that this was due to the electromagnetic interactions with the sea, until Hermann Weyl proved that hole theory was completely symmetric between negative and positive charges. Dirac also predicted a reaction e− + p+ → γ + γ, where an electron and a proton annihilate to give two photons. Robert Oppenheimer and Igor Tamm proved that this would cause ordinary matter to disappear too fast. A year later, in 1931, Dirac modified his theory and postulated the positron, a new particle of the same mass as the electron. The discovery of this particle the next year removed the last two objections to his theory.

However, the problem of infinite charge of the universe remains. Also, as we now know, bosons also have antiparticles, but since bosons do not obey the Pauli exclusion principle (only fermions do), hole theory does not work for them. A unified interpretation of antiparticles is now available in quantum field theory, which solves both these problems.

2.2   Particle-antiparticle annihilation

Main article: Annihilation

If a particle and antiparticle are in the appropriate quantum states, then they can annihilate each other and produce

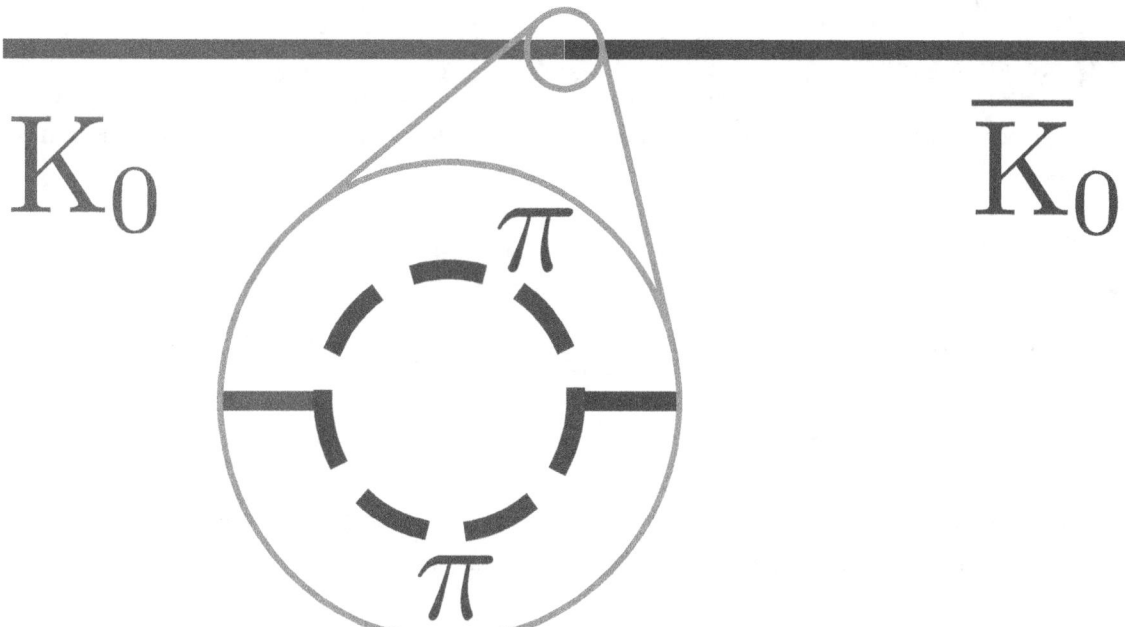

An example of a virtual pion pair that influences the propagation of a kaon, causing a neutral kaon to mix with the antikaon. This is an example of renormalization in quantum field theory— the field theory being necessary because of the change in particle number.

other particles. Reactions such as e− + e+ → γ + γ (the two-photon annihilation of an electron-positron pair) are an example. The single-photon annihilation of an electron-positron pair, e− + e+ → γ, cannot occur in free space because it is impossible to conserve energy and momentum together in this process. However, in the Coulomb field of a nucleus the translational invariance is broken and single-photon annihilation may occur.[4] The reverse reaction (in free space, without an atomic nucleus) is also impossible for this reason. In quantum field theory, this process is allowed only as an intermediate quantum state for times short enough that the violation of energy conservation can be accommodated by the uncertainty principle. This opens the way for virtual pair production or annihilation in which a one particle quantum state may *fluctuate* into a two particle state and back. These processes are important in the vacuum state and renormalization of a quantum field theory. It also opens the way for neutral particle mixing through processes such as the one pictured here, which is a complicated example of mass renormalization.

2.3 Properties of antiparticles

Quantum states of a particle and an antiparticle can be interchanged by applying the charge conjugation (**C**), parity (**P**), and time reversal (**T**) operators. If $|p, \sigma, n\rangle$ denotes the quantum state of a particle (**n**) with momentum **p**, spin **J** whose component in the z-direction is σ, then one has

$$CPT\, |p, \sigma, n\rangle \;=\; (-1)^{J-\sigma}\, |p, -\sigma, n^c\rangle,$$

where **n^c** denotes the charge conjugate state, *i.e.*, the antiparticle. This behaviour under **CPT** is the same as the statement that the particle and its antiparticle lie in the same irreducible representation of the Poincaré group. Properties of antiparticles can be related to those of particles through this. If **T** is a good symmetry of the dynamics, then

$$T\, |p, \sigma, n\rangle \;\propto\; |-p, -\sigma, n\rangle,$$
$$CP\, |p, \sigma, n\rangle \;\propto\; |-p, \sigma, n^c\rangle,$$
$$C\, |p, \sigma, n\rangle \;\propto\; |p, \sigma, n^c\rangle,$$

where the proportionality sign indicates that there might be a phase on the right hand side. In other words, particle and antiparticle must have

- the same mass **m**
- the same spin state **J**
- opposite electric charges **q** and **-q**.

2.4 Quantum field theory

This section draws upon the ideas, language and notation of canonical quantization of a quantum field theory.

One may try to quantize an electron field without mixing the annihilation and creation operators by writing

$$\psi(x) = \sum_k u_k(x) a_k e^{-iE(k)t},$$

where we use the symbol k to denote the quantum numbers p and σ of the previous section and the sign of the energy, $E(k)$, and a_k denotes the corresponding annihilation operators. Of course, since we are dealing with fermions, we have to have the operators satisfy canonical anti-commutation relations. However, if one now writes down the Hamiltonian

$$H = \sum_k E(k) a_k^\dagger a_k,$$

then one sees immediately that the expectation value of H need not be positive. This is because $E(k)$ can have any sign whatsoever, and the combination of creation and annihilation operators has expectation value 1 or 0.

So one has to introduce the charge conjugate *antiparticle* field, with its own creation and annihilation operators satisfying the relations

$$b_{k\prime} = a_k^\dagger \text{ and } b_{k\prime}^\dagger = a_k,$$

where k has the same p, and opposite σ and sign of the energy. Then one can rewrite the field in the form

$$\psi(x) = \sum_{k_+} u_k(x) a_k e^{-iE(k)t} + \sum_{k_-} u_k(x) b_k^\dagger e^{-iE(k)t},$$

where the first sum is over positive energy states and the second over those of negative energy. The energy becomes

$$H = \sum_{k_+} E_k a_k^\dagger a_k + \sum_{k_-} |E(k)| b_k^\dagger b_k + E_0,$$

where E_0 is an infinite negative constant. The vacuum state is defined as the state with no particle or antiparticle, *i.e.*, $a_k|0\rangle = 0$ and $b_k|0\rangle = 0$. Then the energy of the vacuum is exactly E_0. Since all energies are measured relative to the vacuum, **H** is positive definite. Analysis of the properties of *ak* and *bk* shows that one is the annihilation operator for particles and the other for antiparticles. This is the case of a fermion.

This approach is due to Vladimir Fock, Wendell Furry and Robert Oppenheimer. If one quantizes a real scalar field, then one finds that there is only one kind of annihilation operator; therefore, real scalar fields describe neutral bosons. Since complex scalar fields admit two different kinds of annihilation operators, which are related by conjugation, such fields describe charged bosons.

2.4.1   Feynman–Stueckelberg interpretation

By considering the propagation of the negative energy modes of the electron field backward in time, Ernst Stueckelberg reached a pictorial understanding of the fact that the particle and antiparticle have equal mass **m** and spin **J** but opposite charges **q**. This allowed him to rewrite perturbation theory precisely in the form of diagrams. Richard Feynman later gave an independent systematic derivation of these diagrams from a particle formalism, and they are now called Feynman diagrams. Each line of a diagram represents a particle propagating either backward or forward in time. This technique is the most widespread method of computing amplitudes in quantum field theory today.

Since this picture was first developed by Ernst Stueckelberg, and acquired its modern form in Feynman's work, it is called the *Feynman-Stueckelberg interpretation* of antiparticles to honor both scientists.

As a consequence of this interpretation, Villata argued that the assumption of antimatter as CPT-transformed matter would imply that the gravitational interaction between matter and antimatter is repulsive.[5]

2.5   See also

- Gravitational interaction of antimatter

- Parity, charge conjugation and time reversal symmetry.

- CP violations and the baryon asymmetry of the universe.

- Quantum field theory and the list of particles

- Baryogenesis

2.6   References

[1] http://news.nationalgeographic.com/news/2010/11/101118-antimatter-trapped-engines-bombs-nature-science-cern/

[2] Weinberg, Steve. *The quantum theory of fields, Volume 1 : Foundations.* p. 14. ISBN 0-521-55001-7.

[3] Dirac, Paul (1930). "A Theory of Electrons and Protons". *Proceedings of the Royal Society A* **126**(801): 360–365. Bibcode:19 doi:10.1098/rspa.1930.0013.

[4] Sodickson, L.; W. Bowman; J. Stephenson (1961). "Single-Quantum Annihilation of Positrons". *Physical Review* **124** (6): 1851–1861. Bibcode:1961PhRv..124.1851S. doi:10.1103/PhysRev.124.1851.

[5] M. Villata, CPT symmetry and antimatter gravity in general relativity, 2011, EPL (Europhysics Letters) 94, 20001

- Feynman, R. P. (1987). "The reason for antiparticles". In R. P. Feynman and S. Weinberg. *The 1986 Dirac memorial lectures.* Cambridge University Press. ISBN 0-521-34000-4.

- Weinberg, S. (1995). *The Quantum Theory of Fields, Volume 1: Foundations.* Cambridge University Press. ISBN 0-521-55001-7.

Chapter 3

List of Feynman diagrams

This is a list of common Feynman diagrams.

Name or phenomenon	Description	Diagram
Beta decay		
Neutrino-less double beta decay	If neutrinos are Majorana fermions (that is, their own antiparticle), Neutrino-less double beta decay is possible. Several experiments are searching for this.	
Pair creation and annihilation	In the Stückelberg–Feynman interpretation, pair annihilation is the same process as pair creation	
Penguin diagram		
Box diagram	The box diagram for kaon oscillations	
Higgs boson production	Via gluons and top quarks	

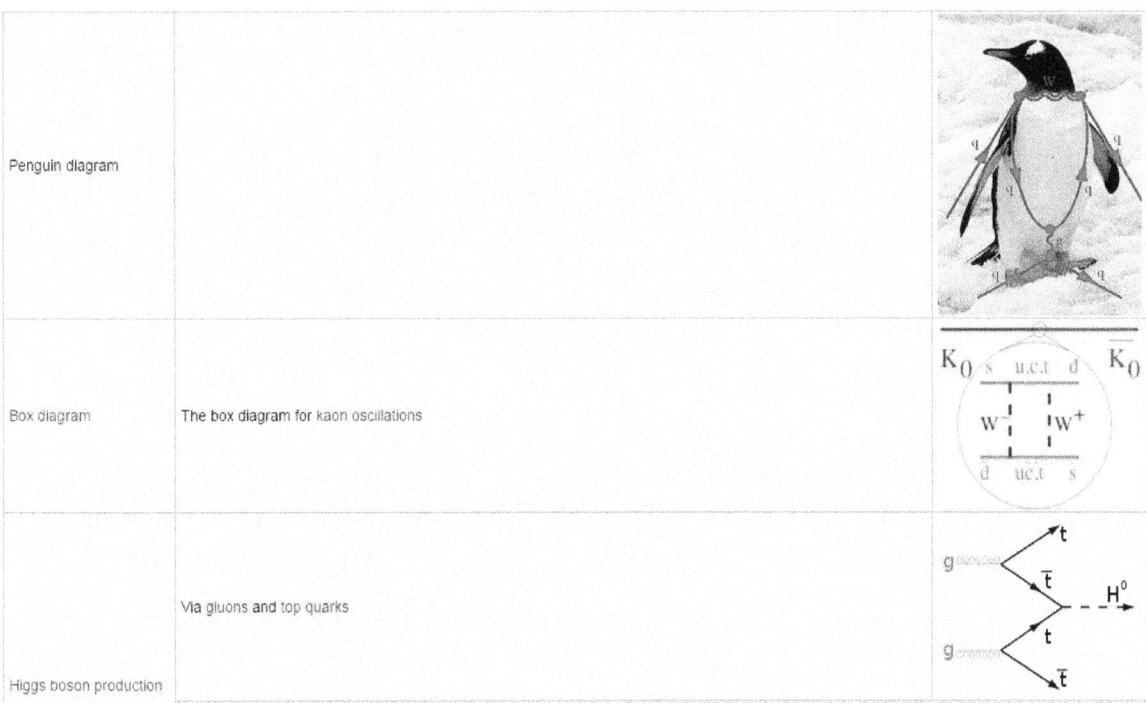

Higgs boson production		
	Via quarks and W or Z bosons	
—	One of the many cancellations to the quadratic divergence to squared mass of the Higgs boson which occurs in the MSSM.	
Primakoff effect		

Chapter 4

Angular momentum diagrams (quantum mechanics)

In quantum mechanics and its applications to quantum many-particle systems, notably quantum chemistry, **angular momentum diagrams**, or more accurately from a mathematical viewpoint **angular momentum graphs**, are a diagrammatic method for representing angular momentum quantum states of a quantum system allowing calculations to be done symbolically. More specifically, the arrows encode angular momentum states in bra–ket notation and include the abstract nature of the state, such as tensor products and transformation rules.

The notation parallels the idea of Penrose graphical notation and Feynman diagrams. The diagrams consist of arrows and vertices with quantum numbers as labels, hence the alternative term "graphs". The sense of each arrow is related to Hermitian conjugation, which roughly corresponds to time reversal of the angular momentum states (c.f. Schrödinger equation). The diagrammatic notation is a considerably large topic in its own right with a number of specialized features – this article introduces the very basics.

They were developed primarily by Adolfas Jucys in the twentieth century.

4.1 Equivalence between Dirac notation and Jucys diagrams

4.1.1 Angular momentum states

The quantum state vector of a single particle with total angular momentum quantum number j and total magnetic quantum number $m = j, j - 1, ..., -j + 1, -j$, is denoted as a ket $|j, m\rangle$. As a diagram this is a *single*headed arrow.

Symmetrically, the corresponding bra is $\langle j, m|$. In diagram form this is a *double*headed arrow, pointing in the opposite direction to the ket.

In each case;

- the quantum numbers j, m are often labelled next to the arrows to refer to a specific angular momentum state,

- arrowheads are almost always placed at the middle of the line, rather than at the tip,

- equals signs "=" are placed between equivalent diagrams, exactly like for multiple algebraic expressions equal to each other.

The most basic diagrams are for kets and bras:

$$j \quad m$$

$=$

$$j \quad m$$

Ket $|j, m\rangle$

$$j \quad m$$

$=$

$$j \quad m$$

Bra $\langle j, m|$

Arrows are directed to or from vertices, a state transforming according to:

- a standard representation is designated by an oriented line leaving a vertex,

- a contrastandard representation is depicted as a line entering a vertex.

As a general rule, the arrows follow each other in the same sense. In the contrastandard representation, the time reversal operator, denoted here by T, is used. It is unitary, which means the Hermitian conjugate $T^{\dagger}$ equals the inverse operator T^{-1}, that is $T^{\dagger} = T^{-1}$. It's action on the position operator leaves it invariant:

$$T\hat{\mathbf{x}}T^{\dagger} = \hat{\mathbf{x}}$$

but the linear momentum operator becomes negative:

$$T\hat{\mathbf{p}}T^{\dagger} = -\hat{\mathbf{p}}$$

and the spin operator becomes negative:

$$T\hat{\mathbf{S}}T^{\dagger} = -\hat{\mathbf{S}}$$

Since the orbital angular momentum operator is $\mathbf{L} = \mathbf{x} \times \mathbf{p}$, this must also become negative:

$$T\hat{\mathbf{L}}T^{\dagger} = -\hat{\mathbf{L}}$$

and therefore the total angular momentum operator $\mathbf{J} = \mathbf{L} + \mathbf{S}$ becomes negative:

$$T\hat{\mathbf{J}}T^{\dagger} = -\hat{\mathbf{J}}$$

Acting on an eigenstate of angular momentum $|j, m\rangle$, it can be shown that [see for example P.E.S. Wormer and J. Paldus (2006)]:

$$T|j, m\rangle \equiv |T(j, m)\rangle = (-1)^{j-m}|j, -m\rangle$$

The time-reversed diagrams for kets and bras are:

$$j \quad m$$

$$= \quad j \quad m$$

Time reversed ket $|j, m\rangle$.

$$j \quad m$$

$$= \quad j \quad m$$

Time reversed bra $\langle j, m|$.

It is important to position the vertex correctly, as forward-time and reversed-time operators would become mixed up.

4.1.2   Inner product

The inner product of two states $|j_1, m_1\rangle$ and $|j_2, m_2\rangle$ is:

$$\langle j_2, m_2 | j_1, m_1 \rangle = \delta_{j_1 j_2} \delta_{m_1 m_2}$$

and the diagrams are:

$$j_2 \; m_2 \qquad j_1 \; m_1$$

$$= \quad j_1 \; m_1 \qquad j_2 \; m_2$$

Inner product of $|j_1, m_1\rangle$ and $|j_2, m_2\rangle$, that is $\langle j_2, m_2 | j_1, m_1 \rangle$.

$$j_2 \; m_2 \qquad j_1 \; m_1$$

$$= \quad j_1 \; m_1 \qquad j_2 \; m_2$$

Time reversed equivalent.

For summations over the inner product, also known in this context as a contraction (c.f. tensor contraction):

$$\sum_m \langle j, m | j, m \rangle = 2j + 1$$

it is conventional to denote the result as a closed circle labelled only by j, not m:

4.1.3   Outer products

The outer product of two states $|j_1, m_1\rangle$ and $|j_2, m_2\rangle$ is an operator:

$$|j_2, m_2\rangle \langle j_1, m_1|$$

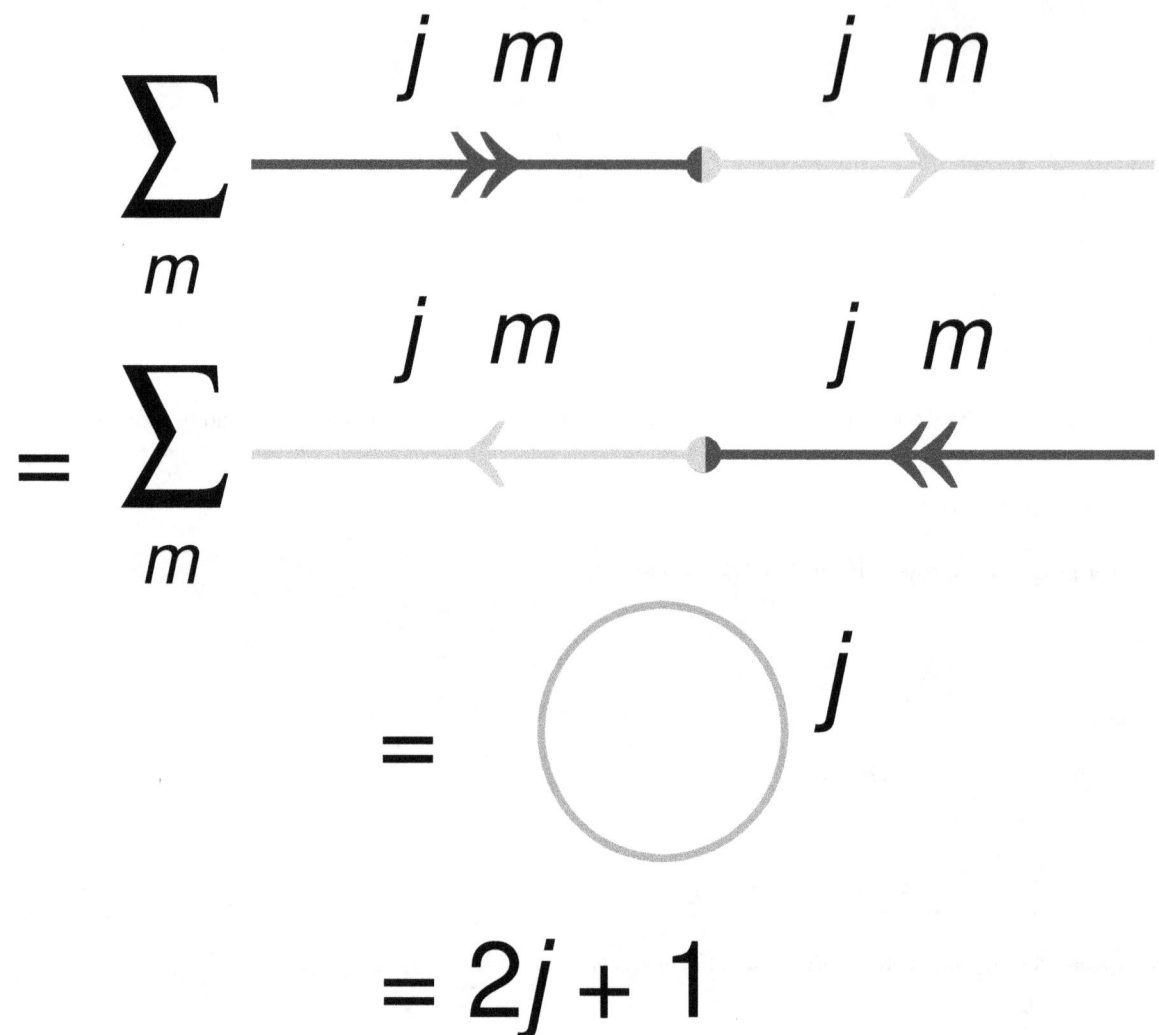

Inner product contraction.

and the diagrams are:

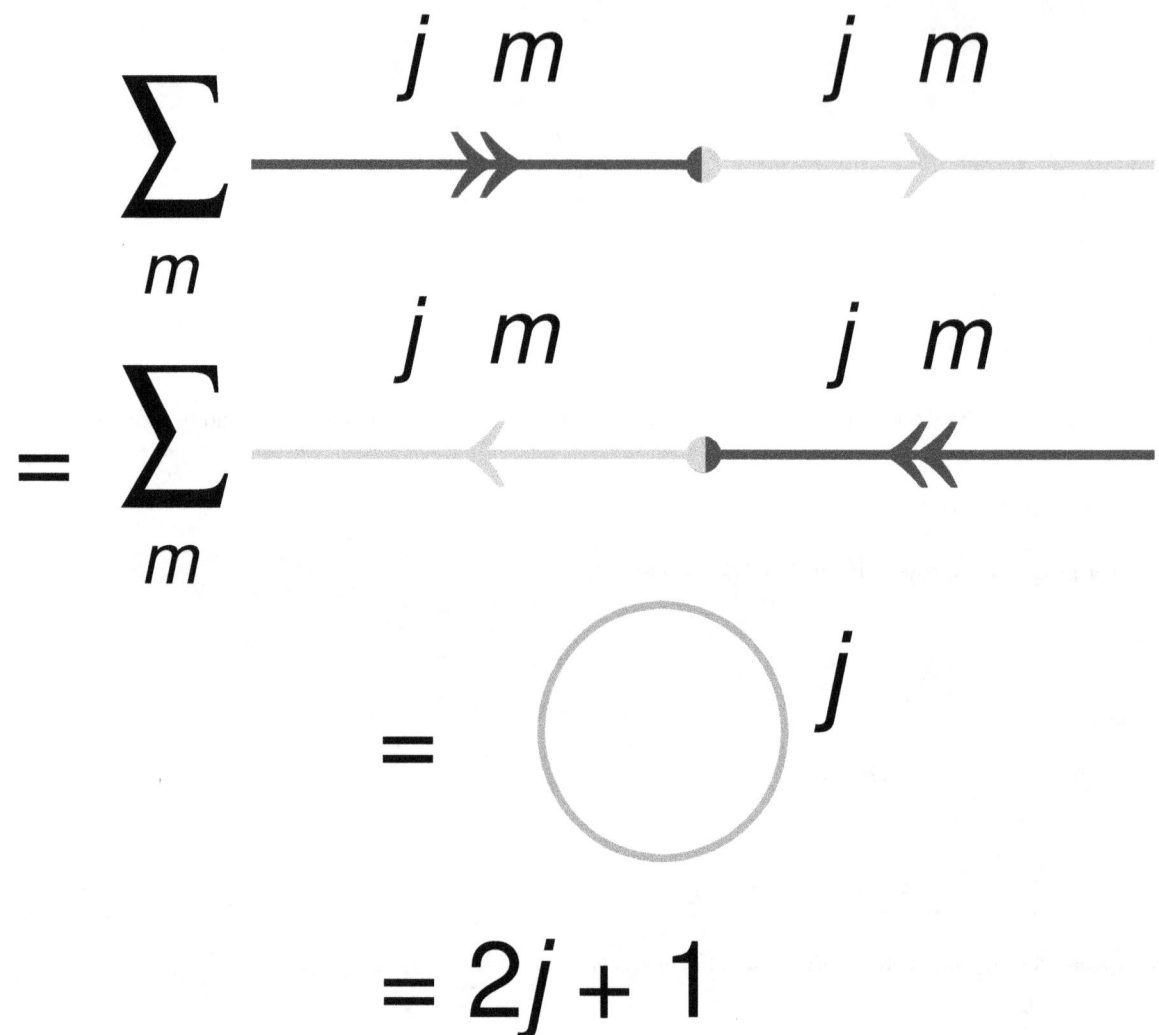

Outer product of $|j_1, m_1\rangle$ and $|j_2, m_2\rangle$, that is $|j_2, m_2\rangle\langle j_1, m_1|$.

Time reversed equivalent.

For summations over the outer product, also known in this context as a contraction (c.f. tensor contraction):

$$\sum_m |j, m\rangle\langle j, m| = \sum_m |j, -m\rangle\langle j, -m|$$

$$= \sum_m (-1)^{2(j-m)} |j, -m\rangle\langle j, -m|$$

$$= \sum_m (-1)^{j-m} |j, -m\rangle\langle j, -m| (-1)^{j-m}$$

$$= \sum_m T|j, m\rangle\langle j, m|T^\dagger$$

where the result for $T|j, m\rangle$ was used, and the fact that m takes the set of values given above. There is no difference between the forward-time and reversed-time states for the outer product contraction, so here they share the same diagram, represented as one line without direction, again labelled by j only and not m:

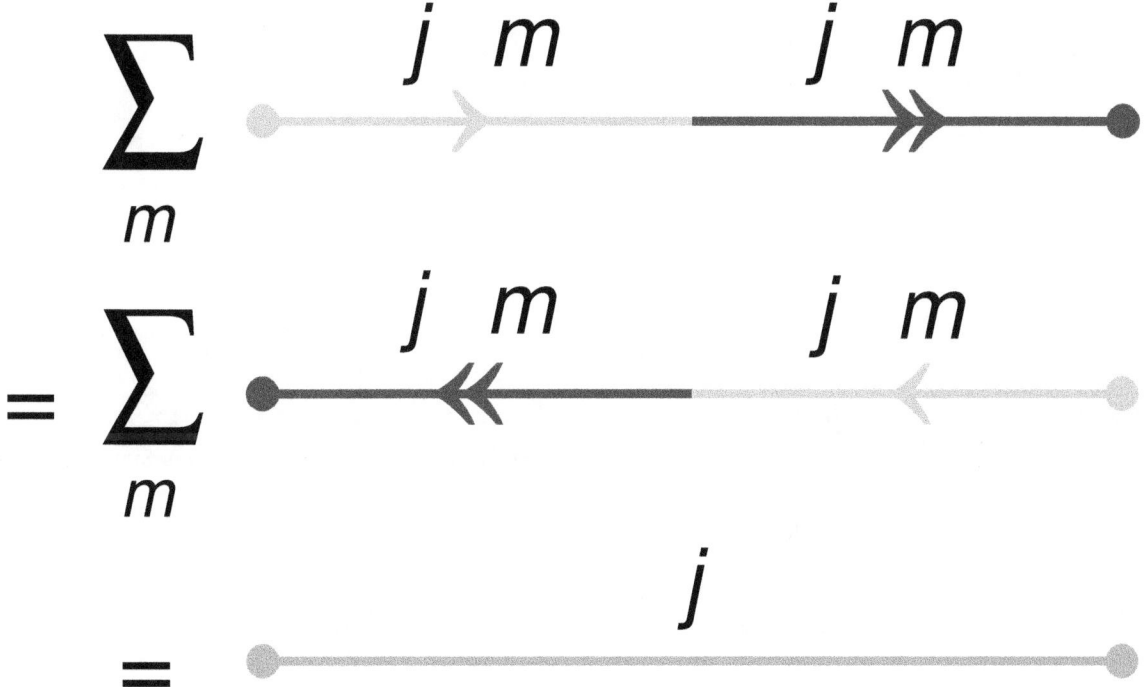

Outer product contraction.

4.1.4 Tensor products

The tensor product $\otimes$ of n states $|j_1, m_1\rangle$, $|j_2, m_2\rangle$, ... $|jn, mn\rangle$ is written

$$|j_1, m_1, j_2, m_2, ...j_n, m_n\rangle \equiv |j_1, m_1\rangle \otimes |j_2, m_2\rangle \otimes \cdots \otimes |j_n, m_n\rangle$$
$$\equiv |j_1, m_1\rangle |j_2, m_2\rangle \cdots |j_n, m_n\rangle$$

and in diagram form, each separate state leaves or enters a common vertex creating a "fan" of arrows - n lines attached to a single vertex.

Vertices in tensor products have signs (sometimes called "node signs"), to indicate the ordering of the tensor-multiplied states:

- a *minus* sign (−) indicates the ordering is *clockwise*, ↻ , and

- a *plus* sign (+) for *anticlockwise*, ↺ .

Signs are of course not required for just one state, diagrammatically one arrow at a vertex. Sometimes curved arrows with the signs are included to show explicitly the sense of tensor multiplication, but usually just the sign is shown with the arrows left out.

Tensor product of $|j_1, m_1\rangle$, $|j_2, m_2\rangle$, $|j_3, m_3\rangle$, that is $|j_1, m_1\rangle|j_2, m_2\rangle|j_3, m_3\rangle = |j_1, m_1, j_2, m_2, j_3, m_3\rangle$. Similarly for more than three angular momenta.

Time reversed equivalent.

For the inner product of two tensor product states:

$$\langle j'_n, m'_n, ..., j'_2, m'_2, j'_1, m'_1 | j_1, m_1, j_2, m_2, ...j_n, m_n \rangle$$
$$= \langle j'_n, m'_n | ... \langle j'_2, m'_2 | \langle j'_1, m'_1 | | j_1, m_1 \rangle | j_2, m_2 \rangle ... | j_n, m_n \rangle$$
$$= \prod_{k=1}^{n} \langle j'_k, m'_k | j_k, m_k \rangle$$

there are n lots of inner product arrows:

Inner product of $|j'_1, m'_1, j'_2, m'_2, j'_3, m'_3\rangle$ and $|j_1, m_1, j_2, m_2, j_3, m_3\rangle$, that is $\langle j'_3, m'_3, j'_2, m'_2, j'_1, m'_1 | j_1, m_1, j_2, m_2, j_3, m_3\rangle$. Similarly for more than three pairs of angular momenta.

Time reversed equivalent.

4.2 Examples and applications

- The diagrams are well-suited for Clebsch–Gordan coefficients.
- Calculations with real quantum systems, such as multielectron atoms and molecular systems.

4.3 See also

- Vector model of the atom
- Ladder operator
- Fock space
- Feynman diagrams

4.4 References

4.4.1 Notes

- P.E.S. Wormer, J. Paldus (2006). "Angular Momentum Diagrams". *Advances In Quantum Chemistry* **51** (Elsevier). pp. 59–124. doi:10.1016/S0065-3276(06)51002-0. ISSN 0065-3276. These authors use the theta variant ϑ for the time reversal operator, here we use T.

- I. Lindgren, J. Morrison (1986). *Atomic Many-Body Theory*. Chemical Physics **13** (2nd ed.). Springer-Verlag. ISBN 3-540-166-491.

4.4.2 Further reading

- G.W.F. Drake (2006). *Springer Handbook of Atomic, Molecular, and Optical Physics* (2nd ed.). springer. p. 60. ISBN 0-3872-6308-X.

- U. Kaldor, S. Wilson (2003). *Theoretical Chemistry and Physics of Heavy and Superheavy Elements*. Progress in Theoretical Chemistry and Physics **11**. springer. p. 183. ISBN 1-4020-1371-X.

- E.J. Brändas, P.O. Löwdin, E. Brändas, E.S. Kryachko (2004). *Fundamental World of Quantum Chemistry: A Tribute to the Memory of Per-Olov Löwdin* **3**. Springer. p. 385. ISBN 1-402-025-831.

- P. Schwerdtfeger (2004). *Relativistic Electronic Structure Theory: Part 2. Applications*. Theoretical and Computational Chemistry **14**. Elsevier. p. 97. ISBN 008-054-047-3.

- M. Barysz, Y. Ishikawa (2010). *Relativistic Methods for Chemists*. Challenges and advances in computational chemistry and physics **10**. Springer. p. 311. ISBN 1-402-099-754.

- G.H.F. Diercksen, S. Wilson (1983). *Methods in Computational Molecular Physics*. Nato Science Series C **113**. Springer. ISBN 9-027-716-382.

- Zenonas Rudzikas (2007). "8". *Theoretical Atomic Spectroscopy*. Cambridge Monographs on Atomic, Molecular and Chemical Physics **7**. University of Chicago: Cambridge University Press. ISBN 0-521-026-229.

- Lietuvos Fizikų draugija (2004). *Lietuvos fizikos žurnalas* **44**. University of Chicago: Draugija.

- P.E.T. Jorgensen (1987). *Operators and Representation Theory: Canonical Models for Algebras of Operators Arising in Quantum Mechanics*. University of Chicago: Elsevier. ISBN 008-087-258-1.

Chapter 5

Minkowski diagram

The **Minkowski diagram**, also known as a **spacetime diagram**, was developed in 1908 by Hermann Minkowski and provides an illustration of the properties of space and time in the special theory of relativity. It allows a quantitative understanding of the corresponding phenomena like time dilation and length contraction without mathematical equations.

The term Minkowski diagram is used in both a generic and particular sense. In general, a Minkowski diagram is a graphic depiction of a portion of Minkowski space, often where space has been curtailed to a single dimension. These two-dimensional diagrams portray worldlines as curves in a plane that correspond to motion along the spatial axis. The vertical axis is usually temporal, and the units of measurement are taken such that the light cone at an event consists of the lines of slope plus or minus one through that event.[1]

A particular Minkowski diagram illustrates the result of a Lorentz transformation. The horizontal corresponds to the usual notion of *simultaneous events*, for a stationary observer at the origin. The Lorentz transformation relates two inertial frames of reference, where an observer makes a change of velocity at the event $(0, 0)$. The new time axis of the observer forms an angle α with the previous time axis, with $\alpha < \pi/4$. After the Lorentz transformation the new simultaneous events lie on a line inclined by α to the previous line of simultaneity. Whatever the magnitude of α, the line $t = x$ forms the universal[2] bisector.

5.1   Basics

For simplification in Minkowski diagrams, usually only events in a universe of one space dimension and one time dimension are considered. Unlike common distance-time diagrams, the distance will be displayed on the horizontal axis and the time on the vertical axis. In this manner the events happening in the one dimension of space can be transferred easily to a horizontal line in the diagram. Objects plotted on the diagram can be thought of as moving from bottom to top as time passes. In this way each object, like an observer or a vehicle, follows in the diagram a certain curve which is called its world line.

Each point in the diagram represents a certain position in space and time. Such a position is called an **event** whether or not anything happens at that position. The space and time units of measurement on the axes may, for example, be taken as one of the following pairs:

- units of 30 centimetres length and nanoseconds, or

- astronomical units and intervals of 8 minutes and 20 seconds (500 seconds), or

- light years and years.

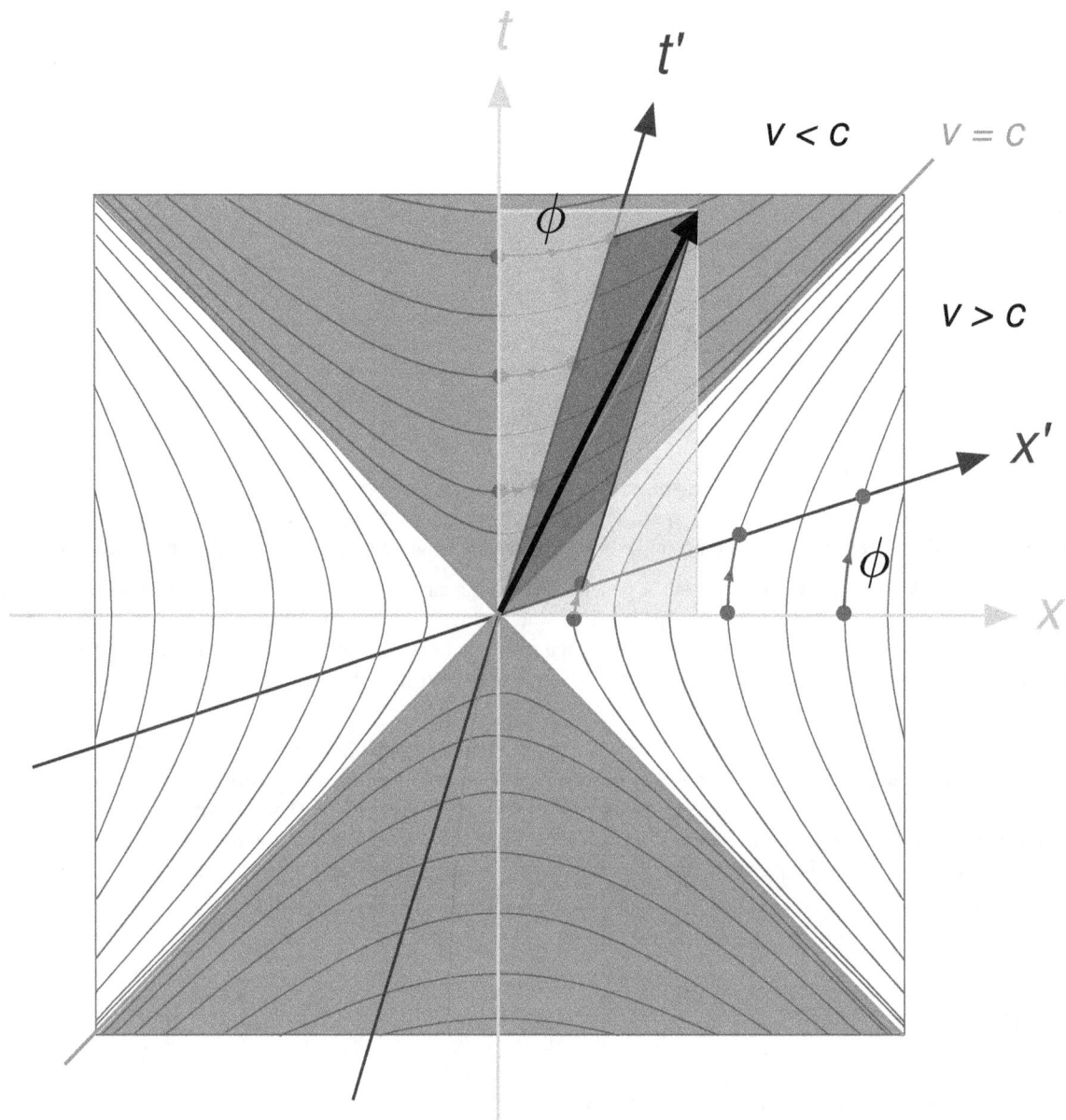

Minkowski diagram with resting frame (x, t), moving frame (x′, t′), light cone, and hyperbolas marking out time and space with respect to the origin.

5.2   Path-time diagram in Newtonian physics

The black axes labelled x and ct on the adjoining diagram are the coordinate system of an observer which we will refer to as 'at rest', and who is positioned at $x = 0$. His world line is identical with the time axis. Each parallel line to this axis would correspond also to an object at rest but at another position. The blue line, however, describes an object moving with constant speed v to the right, such as a moving observer.

This blue line labelled $ct′$ may be interpreted as the time axis for the second observer. Together with the path axis (labeled x, which is identical for both observers) it represents his coordinate system. Both observers agree on the location of the origin of their coordinate systems. The axes for the moving observer are not perpendicular to each other and the scale on his time axis is stretched. To determine the coordinates of a certain event, two lines, each parallel to one of the two axes, must be constructed passing through the event, and their intersections with the axes read off.

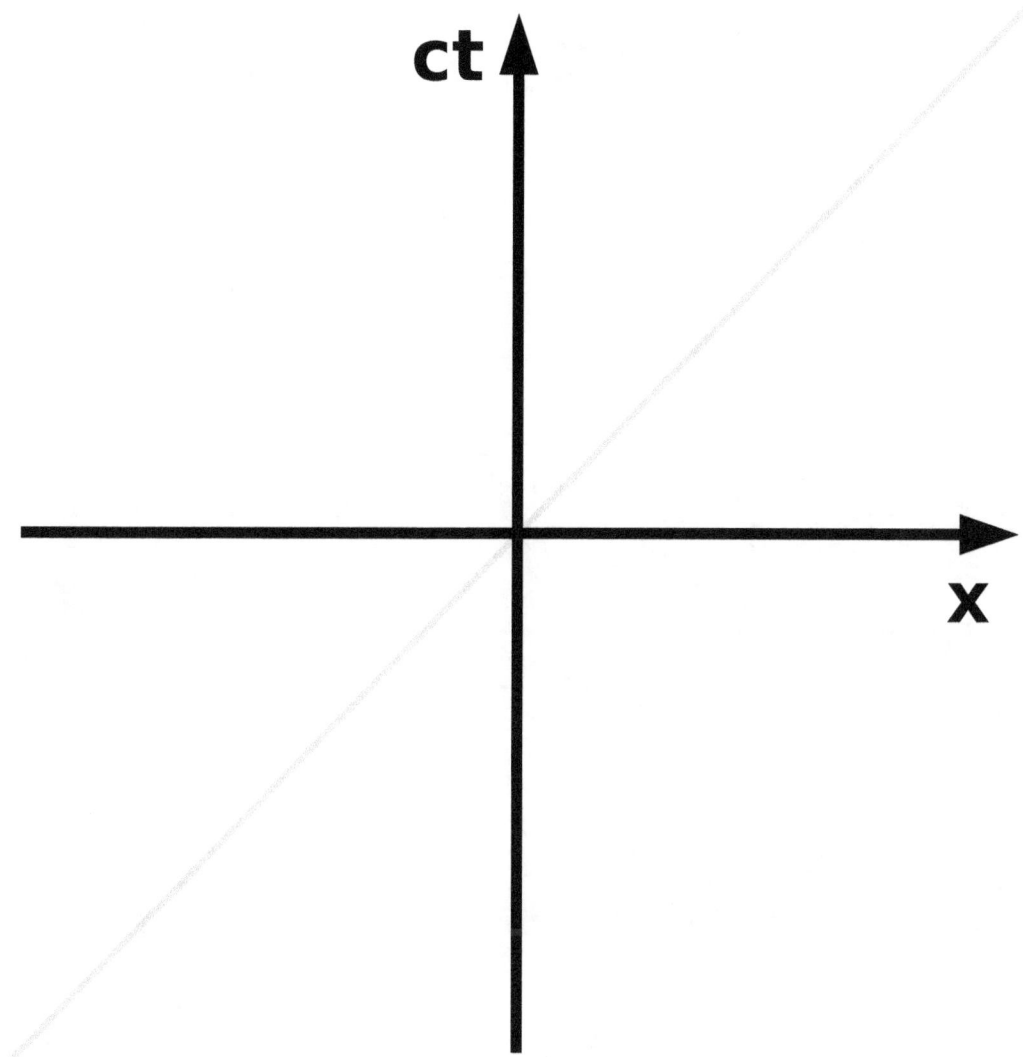

A photon moving right at the origin corresponds to the yellow track of events, a straight line with a slope of 45°.

Determining position and time of the event A as an example in the diagram leads to the same time for both observers, as expected. Only for the position different values result, because the moving observer has approached the position of the event A since $t = 0$. Generally stated, all events on a line parallel to the path axis (x axis) happen simultaneously for both observers. There is only one universal time $t = t'$ which corresponds with the existence of only one common path axis. On the other hand due to two different time axes the observers usually measure different path coordinates for the same event. This graphical translation from x and t to x' and t' and vice versa is described mathematically by the so called Galilean transformation.

5.3 Minkowski diagram in special relativity

Albert Einstein (1905) discovered that the Newtonian description is not correct,[3] with Hermann Minkowski (1908)

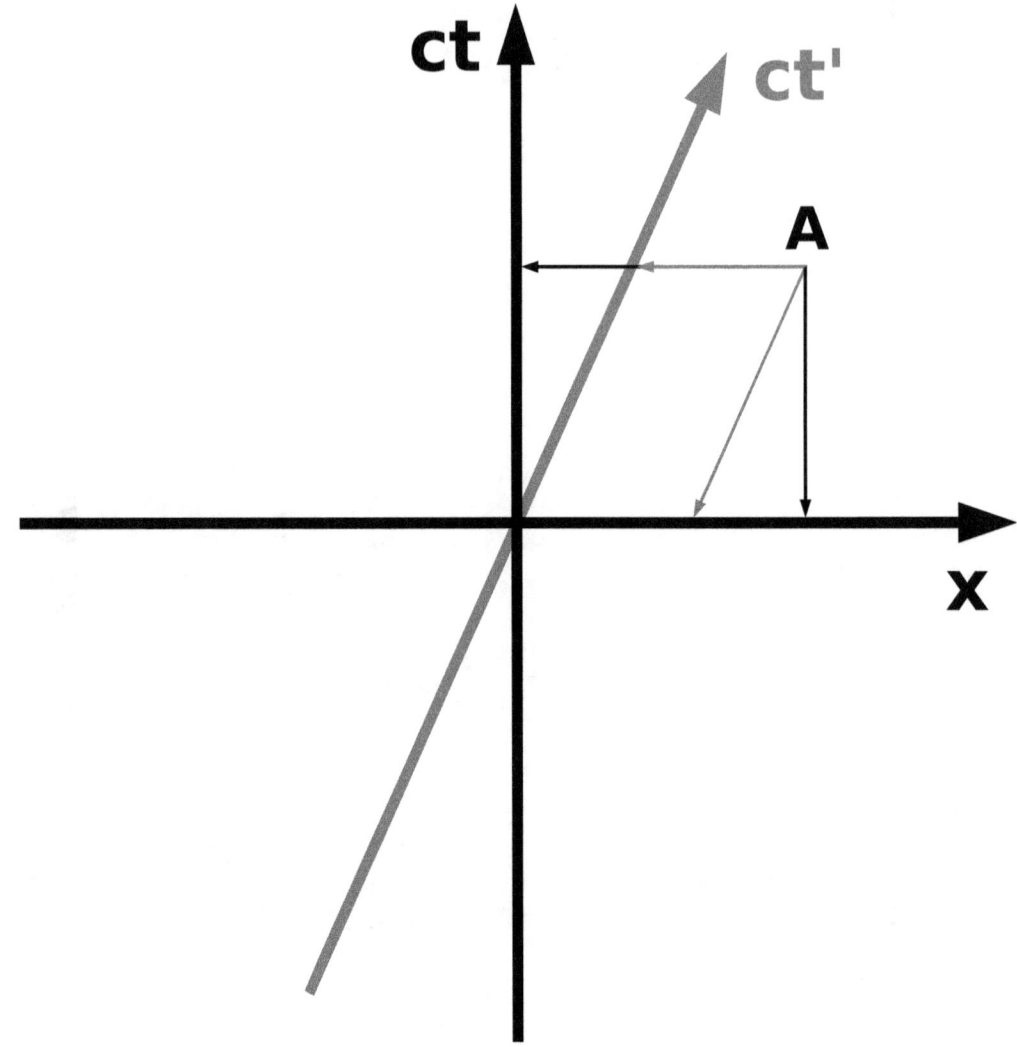

In Newtonian physics for both observers the event at A is assigned to the same point in time.

providing the graphical representation.[4] Space and time have properties which lead to different rules for the translation of coordinates in case of moving observers. In particular, events which are estimated to happen simultaneously from the viewpoint of one observer, happen at different times for the other.

In the Minkowski diagram this relativity of simultaneity corresponds with the introduction of a separate path axis for the moving observer. Following the rule described above each observer interprets all events on a line parallel to his path axis as simultaneous. The sequence of events from the viewpoint of an observer can be illustrated graphically by shifting this line in the diagram from bottom to top.

If *ct* instead of *t* is assigned on the time axes, the angle α between both path axes will be identical with that between both time axes. This follows from the second postulate of the special relativity, saying that the speed of light is the same for all observers, regardless of their relative motion (see below). α is given by

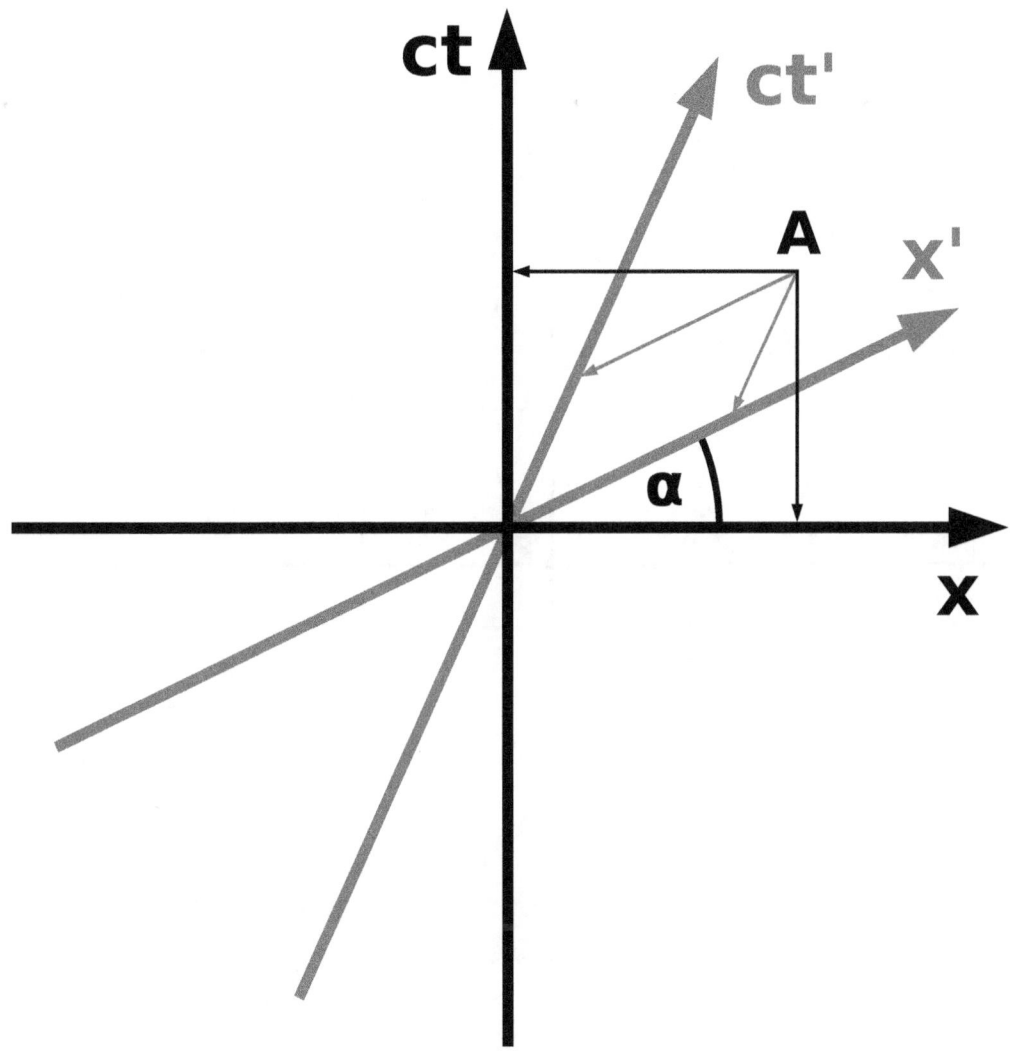

In the theory of relativity each observer assigns the event at A to a different time and location.

$$\tan(\alpha) = \frac{v}{c} = \beta$$

The corresponding translation from x and t to x' and t' and vice versa is described mathematically by the so-called Lorentz transformation. Whatever space and time axes arise through such transformation, in a Minkowski diagram they correspond to conjugate diameters of a pair of hyperbolas. The scales on the axes are given as follows: If U is the unit length on the axes of ct and x respectively, the unit length on the axes of ct' and x' is:[5]

$$U' = U \cdot \sqrt{\frac{1 + \beta^2}{1 - \beta^2}}$$

The ct-axis represents the worldline of a clock resting in S, with U representing the duration between two events happening

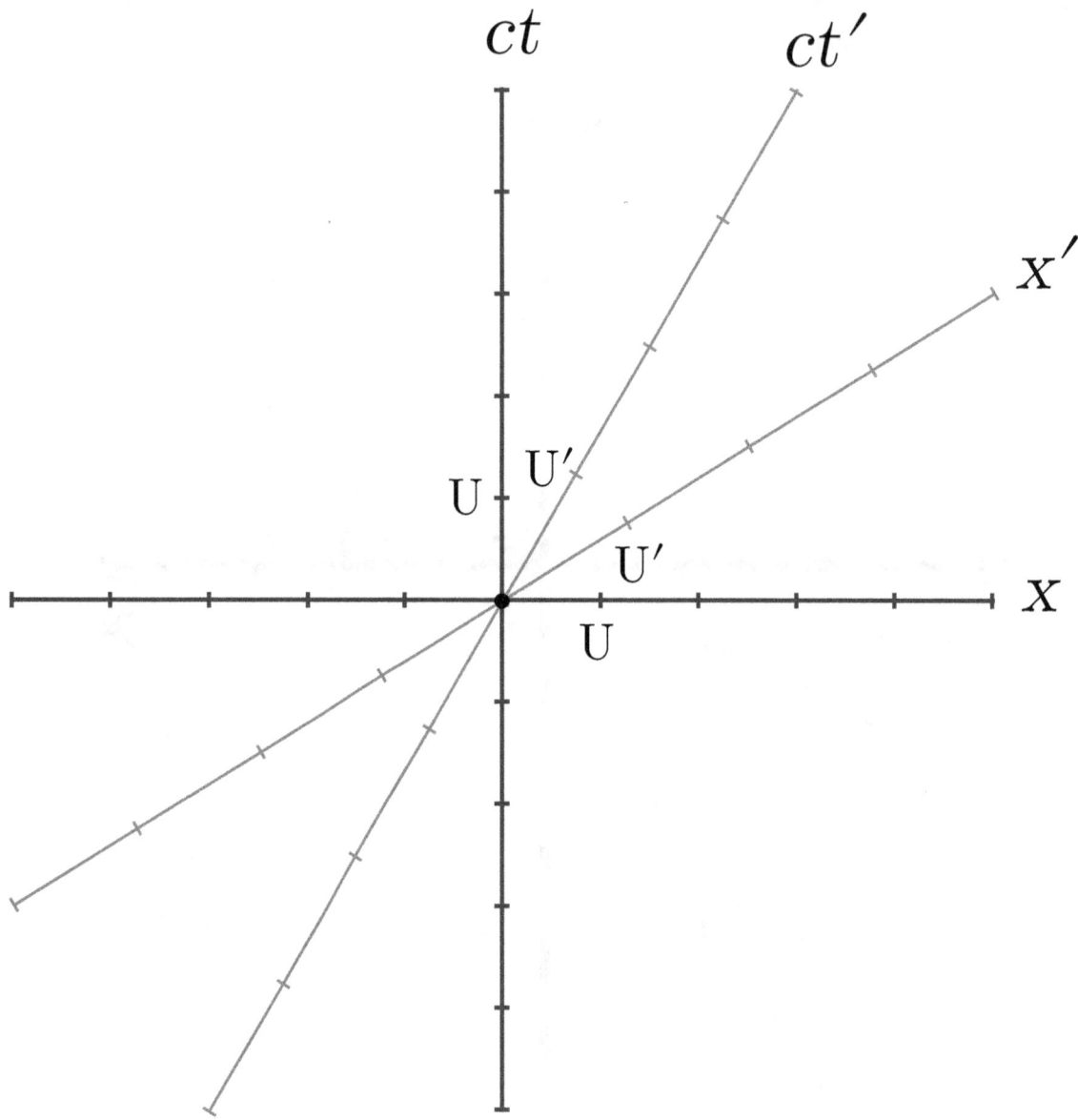

Different scales on the axes.

on this worldline, also called the proper time between these events. Length U upon the x-axis represents the rest length or proper length of a rod resting in S. The same interpretation can also be applied to distance U' upon the ct'- and x'-axis for clocks and rods resting in S'.

History

In Minkowski's 1908 paper there were three diagrams, first to illustrate the Lorentz transformation, then the partition of the plane by the light-cone, and finally illustration of worldlines.[4] The first diagram used a branch of the unit hyperbola $t^2 - x^2 = 1$ to show the locus of a unit of proper time depending on velocity, thus illustrating time dilation. The second diagram showed the conjugate hyperbola to calibrate space, where a similar stretching leaves the impression of FitzGerald contraction. In 1914 Ludwik Silberstein[6] included a diagram of "Minkowski's representation of the Lorentz transformation". This diagram included the unit hyperbola, its conjugate, and a pair of conjugate diameters. Since the 1960s a version of this more complete configuration has been referred to as The Minkowski Diagram, and used as a standard illustration of the transformation geometry of special relativity. E. T. Whittaker has pointed out that the principle of relativity is tantamount to the arbitrariness of what hyperbola radius is selected for time in the Minkowski diagram. In 1912 Gilbert N. Lewis and Edwin B. Wilson applied the methods of synthetic geometry to develop the properties of the non-Euclidean plane that has Minkowski diagrams.[7][8]

5.4 Loedel diagram

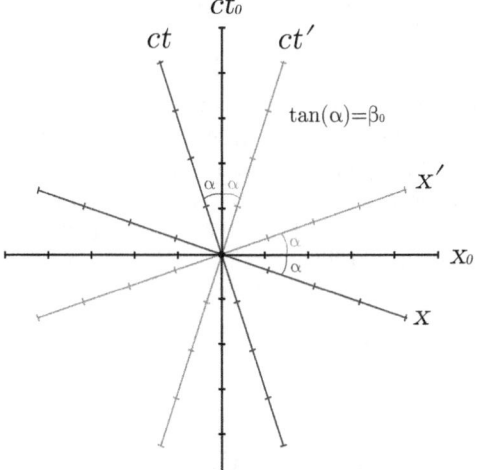

Fig. 1: View in the median frame

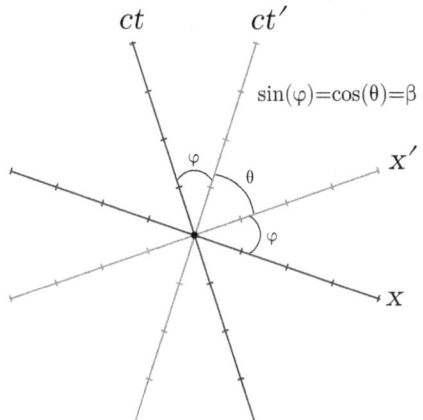

Fig. 2: Symmetric diagram

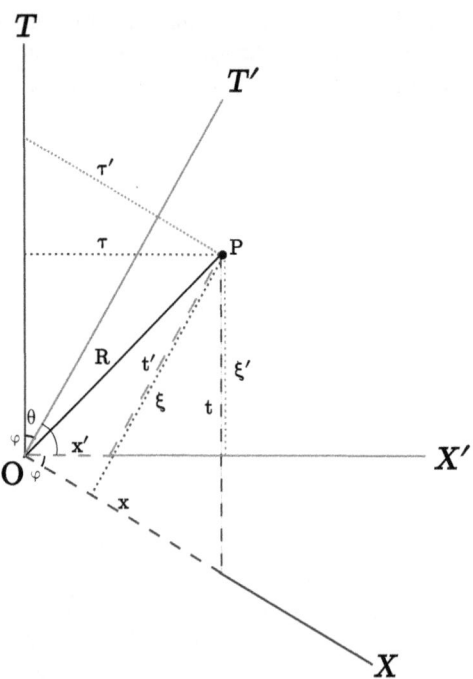

Fig. 3: Covariant and contravariant components.

While the rest frame has space and time axes at right angles, the moving frame has primed axes which form an acute angle. Since the frames are meant to be equivalent, the asymmetry may be disturbing. However, several authors showed that there is a frame of reference between the resting and moving ones where their symmetry would be apparent ("median frame").[9] In this frame, the two other frames are moving in opposite directions with equal speed. Using such coordinates makes the units of length and time the same for both axes. If $\beta = v/c$ and $\gamma = 1/\sqrt{1-\beta^2}$ is given between S and S', then these expressions are connected with the values in their median frame S_0 as follows:[9][10]

(1) $\beta = \dfrac{2\beta_0}{1+\beta_0^2}$,

(2) $\beta_0 = \dfrac{\gamma-1}{\beta\gamma}$.

For instance, if $\beta = 0.5$ between S and S', then by (2) they are moving in their median frame S_0 with approximately $\pm 0.268c$ each in opposite directions. On the other hand, if $\beta_0 = 0.5$ in S_0 , then by (1) the relative velocity between S and S' in their own rest frames is 0.8c. The construction of the axes of S and S' is done in accordance with the ordinary method using $\tan\alpha = \beta_0$ with respect to the orthogonal axes of the median frame (Fig. 1).

However, it turns out that, when drawing such a symmetric diagram, it is possible to derive the diagram's relations even without mentioning the median frame and β_0 at all. Instead, the relative velocity $\beta = v/c$ between S and S' can directly be used in the following construction, providing the same result:[11] If φ is the angle between the axes of ct' and ct (or between x and x'), and θ between the axes of x' and ct', it is given:[11][12][13][14]

$\sin\varphi = \cos\theta = \beta$,

$\cos\varphi = \sin\theta = 1/\gamma$,

$\tan\varphi = \cot\theta = \beta\cdot\gamma$.

Two methods of construction are obvious from Fig. 2: (a) The x-axis is drawn perpendicular to the ct'-axis, the x' and ct-axes are added at angle φ ; (b) the x'-axis is drawn at angle θ with respect to the ct'-axis, the x-axis is added perpendicular to the ct'-axis and the ct-axis perpendicular to the x'-axis.

Also the components of a vector can be vividly demonstrated by such diagrams (Fig. 3): The parallel projections $(x, t; x', t')$ of vector R are its contravariant components, $(\xi, \tau; \xi', \tau')$ its covariant components.[12][13]

History

- Max Born (1920) drew Minkowski diagrams by placing the ct′-axis almost perpendicular to the x-axis, as well as the ct-axis to the x′-axis, in order to demonstrate length contraction and time dilation in the symmetric case of two rods and two clocks moving in opposite direction.[15]

- Dmitry Mirimanoff (1921) showed that there is always a median frame with respect to two relatively moving frames, and derived the relations between them from the Lorentz transformation. However, he didn't give a graphical representation in a diagram.[9]

- Symmetric diagrams were systematically developed by Paul Gruner in collaboration with Josef Sauter in two papers in 1921. Relativistic effects such as length contraction and time dilation and some relations to covariant and contravariant vectors were demonstrated by them.[12][13] Gruner extended this method in subsequent papers (1922-1924), and gave credit to Mirimanoff's treatment as well.[16][17][18][19][20][21]

- The construction of symmetric Minkowski diagrams was later independently rediscovered by several authors. For instance, starting in 1948, Enrique Loedel Palumbo published a series of papers in Spanish language, presenting the details of such an approach.[22][23] In 1955, Henri Amar also published a paper presenting such relations, and gave credit to Loedel in a subsequent paper in 1957.[24][25] Some authors of textbooks use symmetric Minkowski diagrams, denoting as *Loedel diagrams*.[11][14]

5.5   Time dilation

Relativistic time dilation means that a clock (indicating its proper time) that moves relative to an observer is observed to run slower. In fact, time itself in the frame of the moving clock is observed to run slower. This can be read immediately from the adjoining Loedel diagram quite straightforwardly because unit lengths in the two system of axes are identical. Thus, in order to compare reading between the two systems, we can simply compare lengths as they appear on the page: we do not need to consider the fact that unit lengths on each axis are warped by the factor $(1+\beta^2)^{1/2}(1-\beta^2)^{-1/2}$, which we would have to account for in the corresponding Minkowski diagram.

The observer whose reference frame is given by the black axes is assumed to move from the origin O towards A. The moving clock has the reference frame given by the blue axes and moves from O to B. For the black observer, all events happening simultaneously with the event at A are located on a straight line parallel to its space axis. This line passes through A and B, so A and B are simultaneous from the reference frame of the observer with black axes. However, the clock that is moving relative to the black observer marks off time along the blue time axis. This is represented by the distance from O to B. Therefore, the observer at A with the black axes notices his or her clock as reading the distance from O to A while he or she observes the clock moving relative him or her to read the distance from O to B. Due to the distance from O to B being smaller than the distance from O to A, he or she concludes that the time passed on the clock moving relative to him or her is smaller than that passed on his or her own clock.

A second observer, having moved together with the clock from O to B, will argue that the other clock has reached only C until this moment and therefore this clock runs slower. The reason for these apparently paradoxical statements is the different determination of the events happening synchronously at different locations. Due to the principle of relativity, the question of "Who is right?" has no answer and does not make sense.

5.6   Length contraction

Relativistic length contraction means that the proper length of an object moving relative to an observer is decreased and finally also the space itself is contracted in this system. The observer is assumed again to move along the *ct*-axis. The

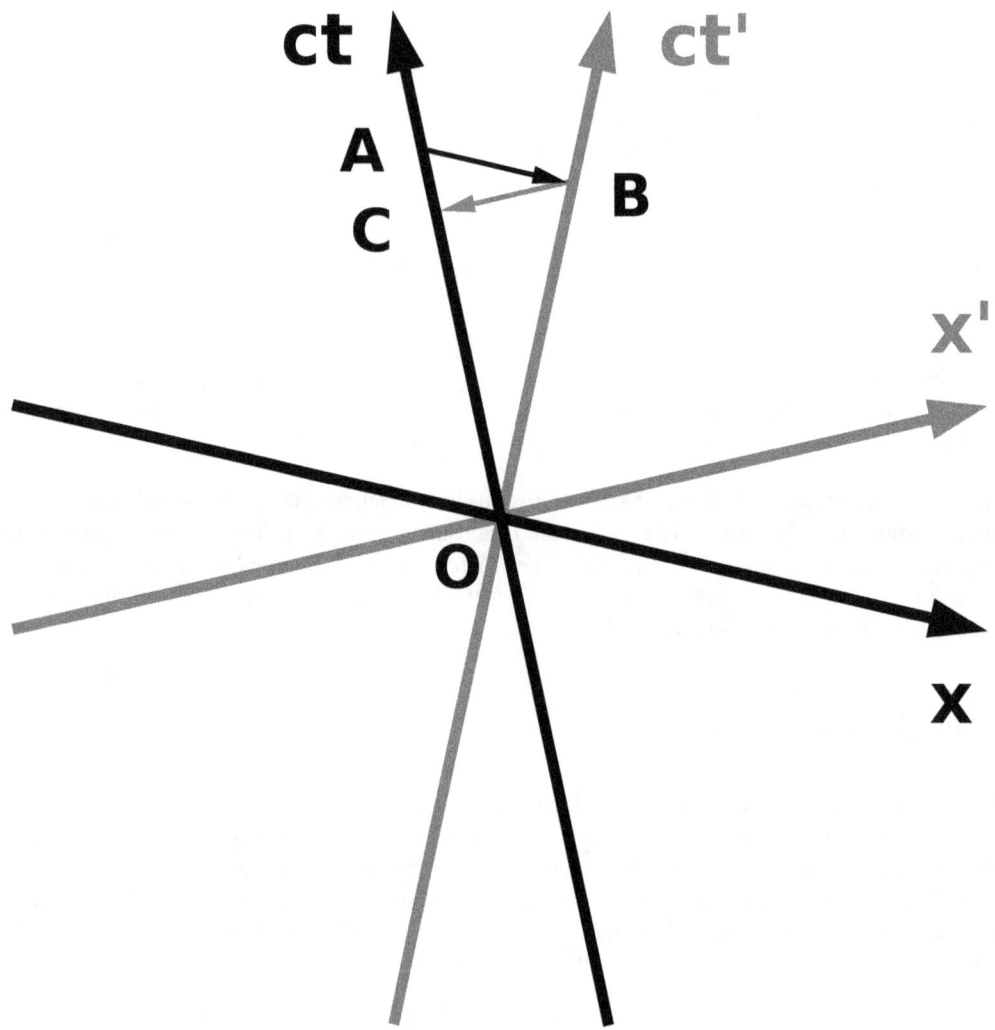

Time dilation: Both observers consider the clock of the other as running slower.

world lines of the endpoints of an object moving relative to him are assumed to move along the ct'-axis and the parallel line passing through A and B. For this observer the endpoints of the object at $t = 0$ are O and A. For a second observer moving together with the object, so that for him the object is at rest, it has the proper length OB at $t' = 0$. Due to OA < OB. the object is contracted for the first observer.

The second observer will argue that the first observer has evaluated the endpoints of the object at O and A respectively and therefore at different times, leading to a wrong result due to his motion in the meantime. If the second observer investigates the length of another object with endpoints moving along the ct-axis and a parallel line passing through C and D he concludes the same way this object to be contracted from OD to OC. Each observer estimates objects moving with the other observer to be contracted. This apparently paradoxical situation is again a consequence of the relativity of simultaneity as demonstrated by the analysis via Minkowski diagram.

For all these considerations it was assumed, that both observers take into account the speed of light and their distance to all events they see in order to determine the actual times at which these events happen from their point of view.

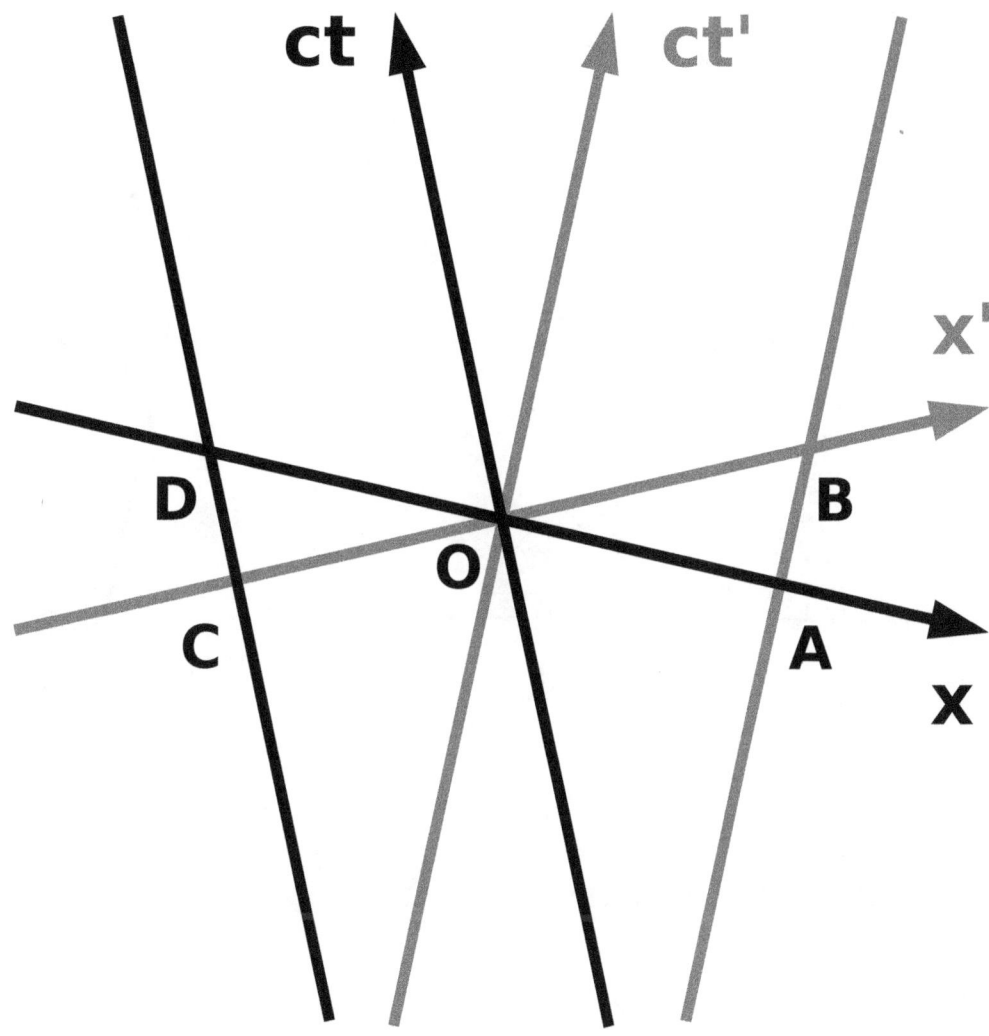

Length contraction: Both observers consider objects moving with the other observer as being shorter.

5.7 Constancy of the speed of light

Another postulate of special relativity is the constancy of the speed of light. It says that any observer in an inertial reference frame measuring the vacuum speed of light relative to himself obtains the same value regardless of his own motion and that of the light source. This statement seems to be paradoxical, but it follows immediately from the differential equation yielding this, and the Minkowski diagram agrees. It explains also the result of the Michelson–Morley experiment which was considered to be a mystery before the theory of relativity was discovered, when photons were thought to be waves through an undetectable medium.

For world lines of photons passing the origin in different directions $x = ct$ and $x = -ct$ holds. That means any position

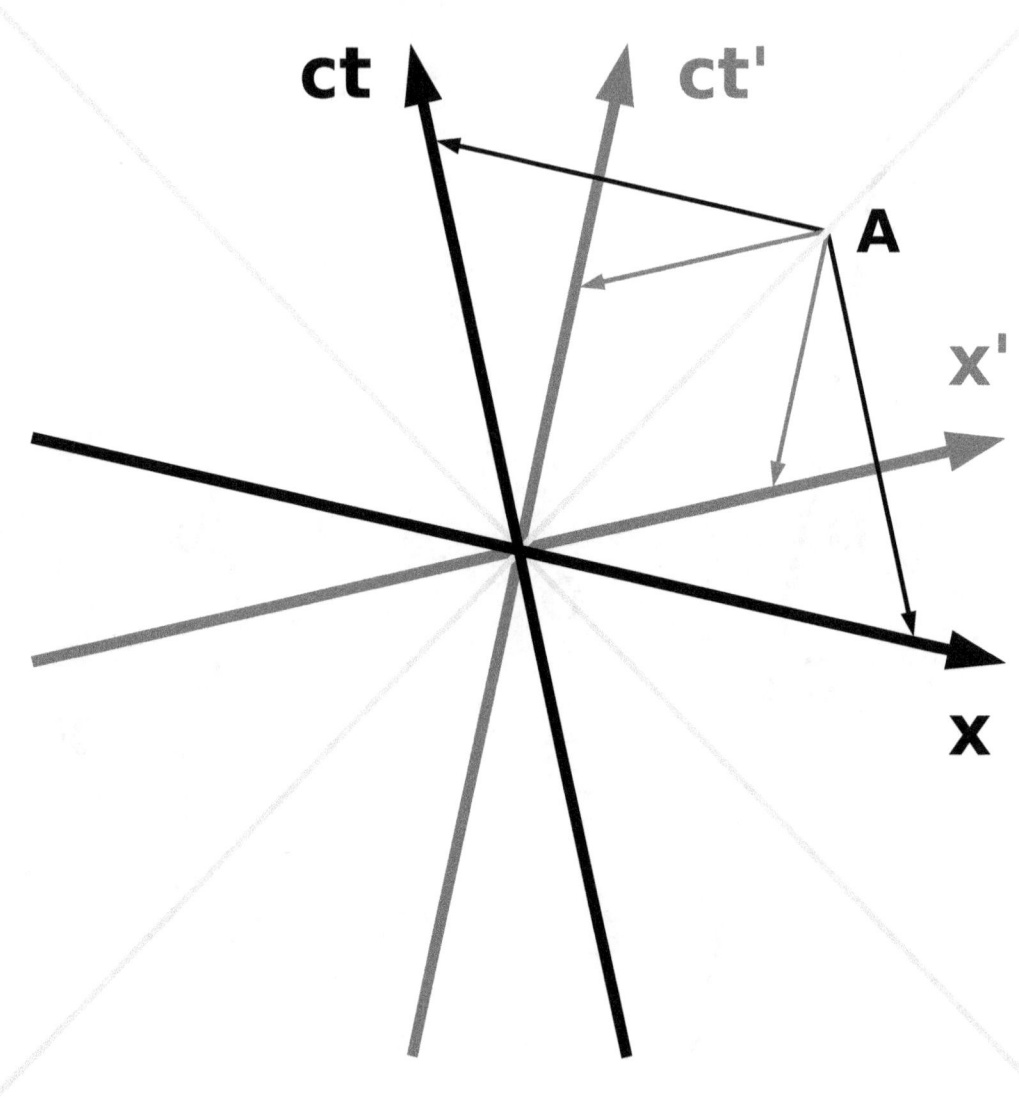

For the speed of a photon passing A both observers measure the same value even though they move relative to each other.

on such a world line corresponds with steps on *x*- and *ct*-axis of equal absolute value. From the rule for reading off coordinates in coordinate system with tilted axes follows that the two world lines are the angle bisectors of the *x*- and *ct*-axis. The Minkowski diagram shows, that they are angle bisectors of the *x′*- and *ct′*-axis as well. That means both observers measure the same speed *c* for both photons.

Further coordinate systems corresponding to observers with arbitrary velocities can be added to this Minkowski diagram. For all these systems both photon world lines represent the angle bisectors of the axes. The more the relative speed approaches the speed of light the more the axes approach the corresponding angle bisector. The path axis is always more flat and the time axis more steep than the photon world lines. The scales on both axes are always identical, but usually different from those of the other coordinate systems.

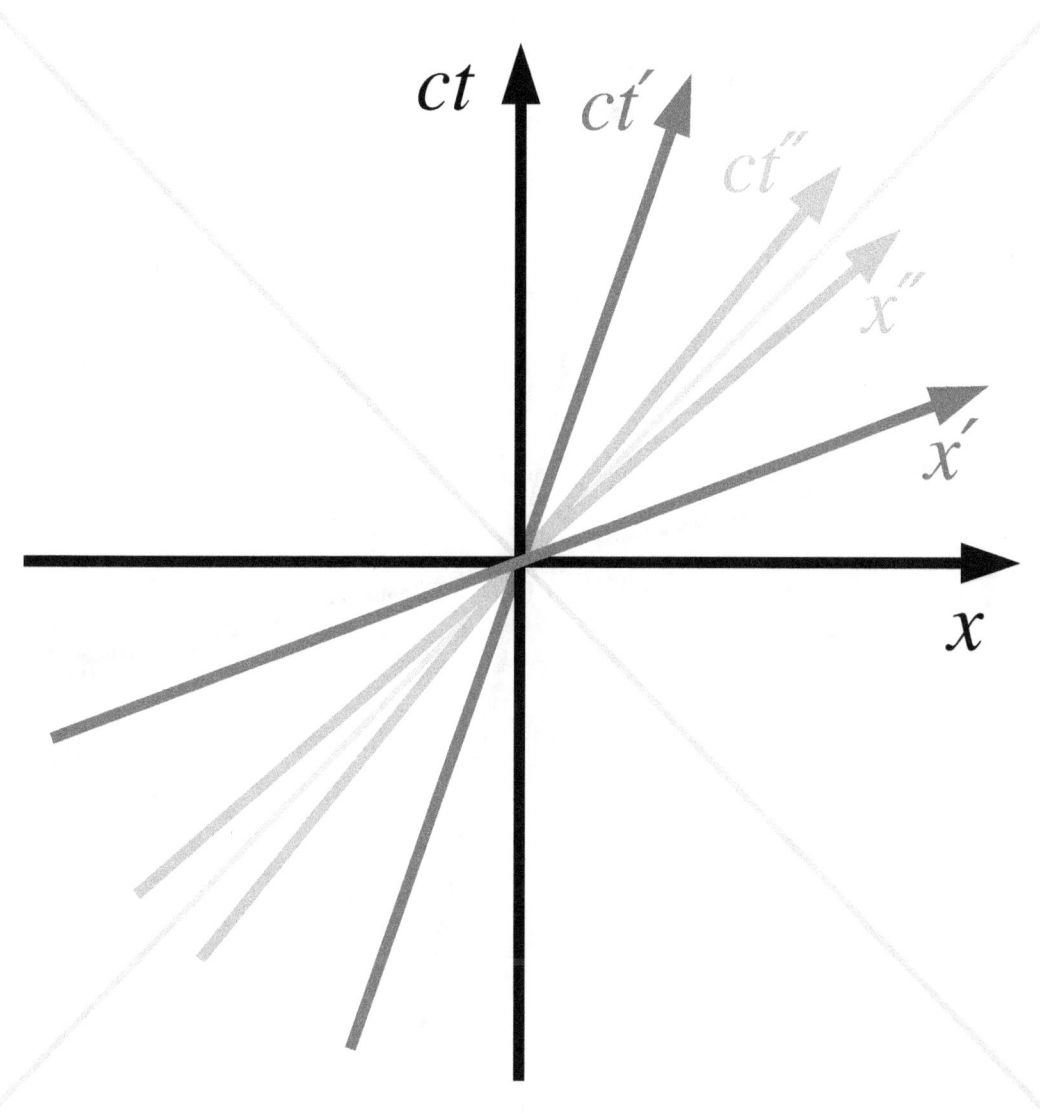

Minkowski diagram for 3 coordinate systems. For the speeds relative to the system in black v′ = 0.4c and v″ = 0.8c holds.

5.8 Speed of light and causality

Straight lines passing the origin which are steeper than both photon world lines correspond with objects moving more slowly than the speed of light. If this applies to an object, then it applies from the viewpoint of all observers, because the world lines of these photons are the angle bisectors for any inertial reference frame. Therefore any point above the origin and between the world lines of both photons can be reached with a speed smaller than that of the light and can have a cause-effect-relationship with the origin. This area is the absolute future, because any event there happens later compared to the event represented by the origin regardless of the observer, which is obvious graphically from the Minkowski diagram.

Following the same argument the range below the origin and between the photon world lines is the absolute past relative to the origin. Any event there belongs definitely to the past and can be the cause of an effect at the origin.

The relationship between any such pairs of event is called *timelike*, because they have a time distance greater than zero for all observers. A straight line connecting these two events is always the time axis of a possible observer for whom they

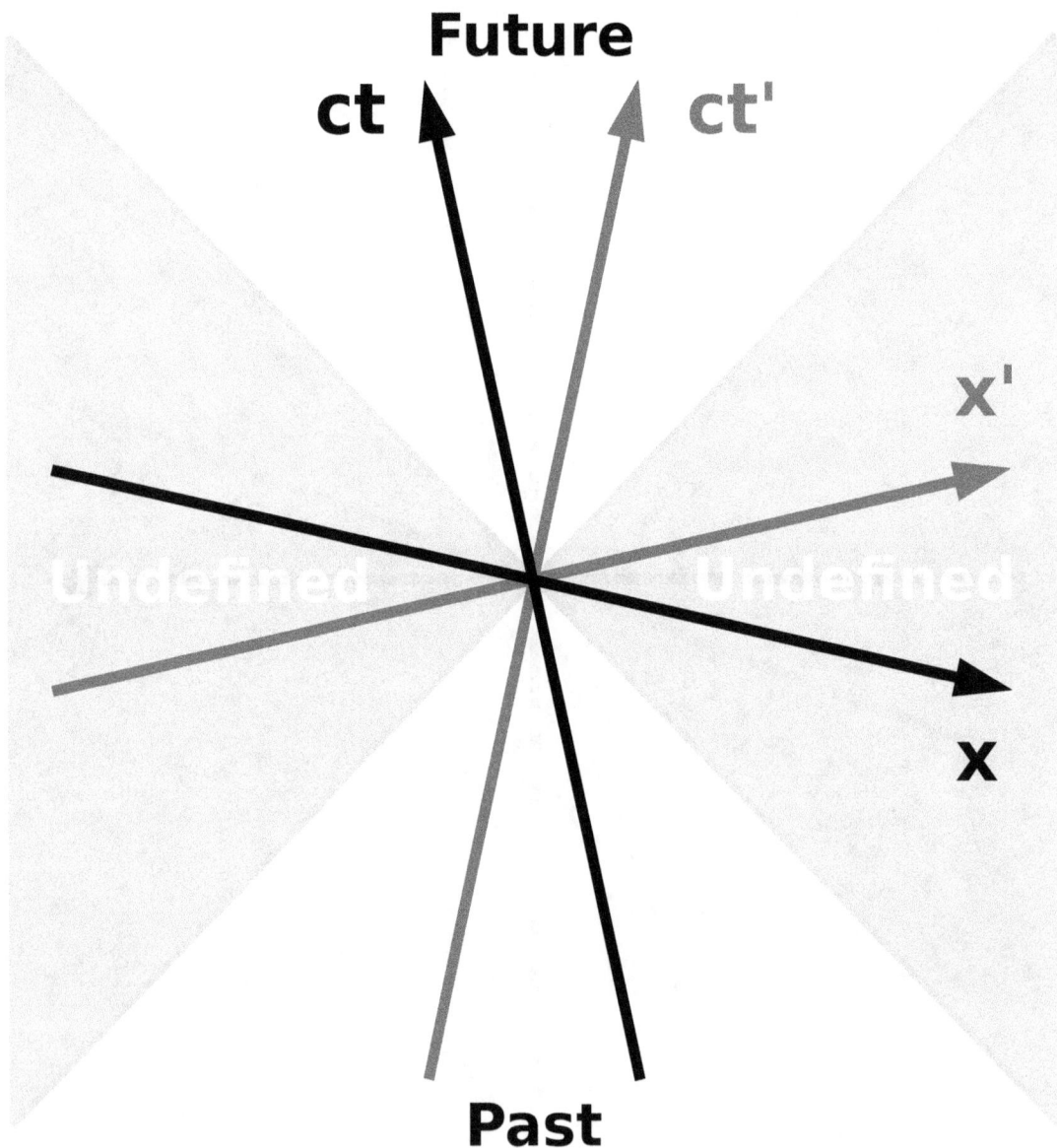

Past and future relative to the origin. For the grey areas a corresponding temporal classification is not possible.

happen at the same place. Two events which can be connected just with the speed of light are called *lightlike*.

In principle a further dimension of space can be added to the Minkowski diagram leading to a three-dimensional representation. In this case the ranges of future and past become cones with apexes touching each other at the origin. They are called light cones.

5.9   The speed of light as a limit

Following the same argument, all straight lines passing through the origin and which are more nearly horizontal than the photon world lines, would correspond to objects or signals moving faster than light regardless of the speed of the observer. Therefore no event outside the light cones can be reached from the origin, even by a light-signal, nor by any object or

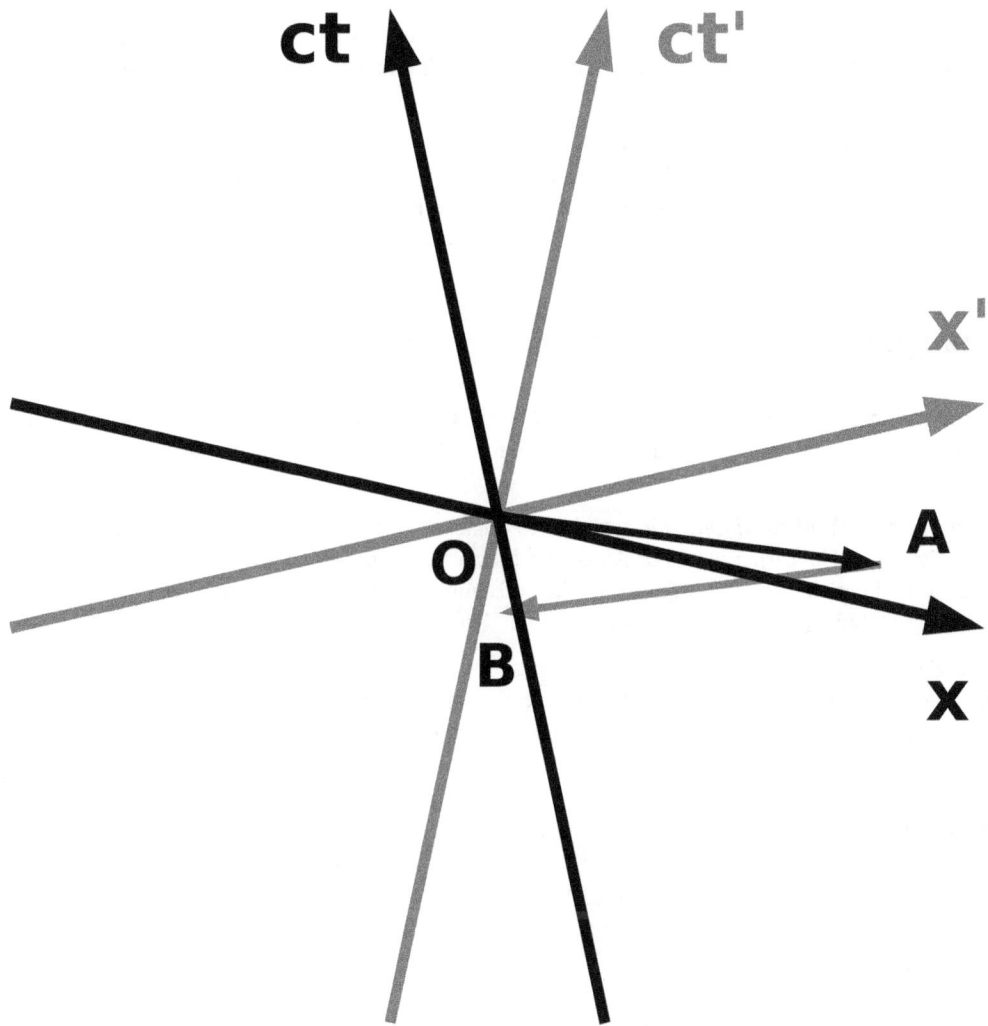

Sending a message at superluminal speed from O via A to B into the past. Both observers consider the temporal order of the pairs of events O and A as well as A and B different.

signal moving with less than the speed of light. Such pairs of events are called *spacelike* because they have a finite spatial distance different from zero for all observers. On the other hand a straight line connecting such events is always the space coordinate axis of a possible observer for whom they happen at the same time. By a slight variation of the velocity of this coordinate system in both directions it is always possible to find two inertial reference frames whose observers estimate the chronological order of these events to be different.

Therefore an object moving faster than light, say from O to A in the adjoining diagram, would imply that, for any observer watching the object moving from O to A, there can be found another observer (moving at less than the speed of light with respect to the first) for whom the object moves from A to O. The question of which observer is right has no unique answer, and therefore makes no physical sense. Any such moving object or signal would violate the principle of causality.

Also, any general technical means of sending signals faster than light would permit information to be sent into the originator's own past. In the diagram, an observer at O in the *x-ct*-system sends a message moving faster than light to A. At A it is received by another observer, moving so as to be in the x'-ct'-system, who sends it back, again faster than light by the same technology, arriving at B. But B is in the past relative to O. The absurdity of this process becomes obvious when both observers subsequently confirm that they received no message at all but all messages were directed towards the other observer as can be seen graphically in the Minkowski diagram. Indeed, if it were possible to accelerate an observer to the speed of light, then the space and time axes would coincide with their angle bisector. The coordinate system would collapse.

These considerations show that the speed of light as a limit is a consequence of the properties of spacetime, and not of the properties of objects such as technologically imperfect space ships. The prohibition of faster-than-light motion actually has nothing in particular to do with electromagnetic waves or light, but depends on the structure of spacetime.

5.10   Eponym

When Taylor and Wheeler composed *Spacetime Physics* (1966), they did *not* use the term "Minkowski diagram" for their spacetime geometry. Instead they included an acknowledgement of Minkowski's contribution to philosophy by the totality of his innovation of 1908.[26]

As an eponym, the term *Minkowski diagram* is subject to Stigler's law of eponymy, namely that Minkowski is wrongly designated as originator. The earlier works of Alexander Macfarlane contain algebra and diagrams that correspond well with the Minkowski diagram. See for instance the plate of figures in *Proceedings of the Royal Society in Edinburgh* for 1900. Macfarlane was building on what one sees in William Kingdon Clifford's *Elements of Dynamic* (1878), page 90.

When abstracted to a line drawing, then any figure showing conjugate hyperbolas, with a selection of conjugate diameters, falls into this category. Students making drawings to accompany the exercises in George Salmon's *A Treatise on Conic Sections* (1900) at pages 165–71 (on conjugate diameters) will be making Minkowski diagrams.

5.11   See also

- Minkowski space
- Penrose diagram
- Rapidity

5.12   References

[1] Mermin (1968) Chapter 17

[2] See Vladimir Karapetoff

[3] Einstein, Albert (1905a), "Zur Elektrodynamik bewegter Körper" (PDF), *Annalen der Physik* **322**(10):891–921, Bibcode: doi:10.1002/andp.19053221004. See also: English translation.

[4] Minkowski, Hermann (1909), "Raum und Zeit", *Physikalische Zeitschrift* **10**: 75–88

   - Various English translations on Wikisource: Space and Time

[5] Jürgen Freund (2008). *Special Relativity for Beginners: A Textbook for Undergraduates*. World Scientific. p. 49. ISBN 981277159X.

[6] Silberstein (1914) *The Theory of Relativity*, page 131

[7] Edwin B. Wilson & Gilbert N. Lewis (1912). "The Space-time Manifold of Relativity. The Non-Euclidean Geometry of Mechanics and Electromagnetics", Proceedings of the American Academy of Arts and Sciences 48:387–507

[8] Synthetic Spacetime, a digest of the axioms used, and theorems proved, by Wilson and Lewis. Archived by WebCite

[9] Mirimanoff, Dmitry (1921). "La transformation de Lorentz-Einstein et le temps universel de M. Ed. Guillaume". *Archives des sciences physiques et naturelles (supplement)*. 5 **3**: 46–48. (Translation: The Lorentz-Einstein transformation and the universal time of Ed. Guillaume)

[10] Albert Shadowitz (2012). *The Electromagnetic Field* (Reprint of 1975 ed.). Courier Dover Publications. p. 460. ISBN 0486132013. See Google books, p. 460

[11] Leo Sartori (1996). *Understanding Relativity: a simplified approach to Einstein's theories*. University of California Press. pp. 151ff. ISBN 0-520-20029-2.

[12] Gruner, Paul & Sauter, Josef (1921). "Représentation géométrique élémentaire des formules de la théorie de la relativité". *Archives des sciences physiques et naturelles*. 5 **3**: 295–296. (Translation: Elementary geometric representation of the formulas of the special theory of relativity)

[13] Gruner, Paul (1921). "Eine elementare geometrische Darstellung der Transformationsformeln der speziellen Relativitätstheorie". *Physikalische Zeitschrift* **22**: 384–385. (Translation: An elementary geometrical representation of the transformation formulas of the special theory of relativity)

[14] Albert Shadowitz (1988). *Special relativity* (Reprint of 1968 ed.). Courier Dover Publications. pp. 20–22. ISBN 0-486-65743-4.

[15] Born, Max (1920). *Die Relativitätstheorie Einsteins* (First ed.). Springer. pp. 177–180. See also Reprint (2013) of third edition (1922) at Google books, p. 187

[16] Gruner, Paul (1922). *Elemente der Relativitätstheorie* [*Elements of the theory of relativity*]. Bern: P. Haupt.

[17] Gruner, Paul (1922). "Graphische Darstellung der speziellen Relativitätstheorie in der vierdimensionalen Raum-Zeit-Welt I" [Graphical representation of the special theory of relativity in the four-dimensional spacetime-world I]. *Zeitschrift für Physik* **10** (1): 22–37. Bibcode:1922ZPhy...10...22G. doi:10.1007/BF01332542.

[18] Gruner, Paul (1922). "Graphische Darstellung der speziellen Relativitätstheorie in der vierdimensionalen Raum-Zeit-Welt II" [Graphical representation of the special theory of relativity in the four-dimensional spacetime-world II]. *Zeitschrift für Physik* **10** (1): 227–235. Bibcode:1922ZPhy...10..227G. doi:10.1007/BF01332563.

[19] Gruner, Paul (1921). "a) Représentation graphique de l'univers espace-temps à quatre dimensions. b) Représentation graphique du temps universel dans la théorie de la relativité". *Archives des sciences physiques et naturelles*. 5 **4**: 234–236. (Translation: Graphical representation of the four-dimensional space-time universe)

[20] Gruner, Paul (1922). "Die Bedeutung "reduzierter" orthogonaler Koordinatensysteme für die Tensoranalysis und die spezielle Relativitätstheorie" [The importance of "reduced" orthogonal coordinate-systems for tensor analysis and the special theory of relativity]. *Zeitschrift für Physik* **10** (1): 236–242. Bibcode:1922ZPhy...10..236G. doi:10.1007/BF01332564.

[21] Gruner, Paul (1924). "Geometrische Darstellungen der speziellen Relativitätstheorie, insbesondere des elektromagnetischen Feldes bewegter Körper" [Geometrich representations of the special theory of relativity, in particular the electromagnetic field of moving bodies]. *Zeitschrift für Physik* **21** (1): 366–371. Bibcode:1924ZPhy...21..366G. doi:10.1007/BF01328285.

[22] Loedel, Enrique (1948). "Aberracion y Relatividad". *Anales soc. cient. argentina* **145**: 3–13.

[23] *Fisica relativista*, Kapelusz Editorial, Buenos Aires, Argentina (1955).

[24] Amar, Henri (1955). "New Geometric Representation of the Lorentz Transformation". *American Journal of Physics* **23** (8): 487–489. Bibcode:1955AmJPh..23..487A. doi:10.1119/1.1934074.

[25] Amar, Henri & Loedel, Enrique (1957). "Geometric Representation of the Lorentz Transformation". *American Journal of Physics* **25** (5): 326–327. Bibcode:1957AmJPh..25..326A. doi:10.1119/1.1934453.

[26] Taylor/Wheeler (1966) page 37: "Minkowski's insight is central to the understanding of the physical world. It focuses attention on those quantities, such as interval, which are the same in all frames of reference. It brings out the relative character of quantities, such as velocity, energy, time, distance, which depend on the frame of reference."

- Anthony French (1968) *Special Relativity*, pages 82 & 83, New York: W W Norton & Company.

- E.N. Glass (1975) "Lorentz boosts and Minkowski diagrams" American Journal of Physics 43:1013,4.

- N. David Mermin (1968) *Space and Time in Special Relativity*, Chapter 17 Minkowski diagrams: The Geometry of Spacetime, pages 155–99 McGraw-Hill.

- Rindler, Wolfgang (2001). *Relativity: Special, General and Cosmological*. Oxford University Press. ISBN 0-19-850836-0.

- W.G.V. Rosser (1964) *An Introduction to the Theory of Relativity*, page 256, Figure 6.4, London: Butterworths.

- Edwin F. Taylor and John Archibald Wheeler (1963) *Spacetime Physics*, pages 27 to 38, New York: W. H. Freeman and Company, Second edition (1992).

- Walter, Scott (1999), "The non-Euclidean style of Minkowskian relativity" (PDF), in J. Gray, *The Symbolic Universe: Geometry and Physics*, Oxford University Press, pp. 91–127 (see page 10 of e-link)

5.13   External links

Media related to Minkowski diagrams at Wikimedia Commons

Chapter 6

Bubble chamber

A **bubble chamber** is a vessel filled with a superheated transparent liquid (most often liquid hydrogen) used to detect electrically charged particles moving through it. It was invented in 1952 by Donald A. Glaser,[1] for which he was awarded the 1960 Nobel Prize in Physics.[2] Supposedly, Glaser was inspired by the bubbles in a glass of beer; however, in a 2006 talk, he refuted this story, although saying that while beer was not the inspiration for the bubble chamber, he did experiments using beer to fill early prototypes.[3]

Cloud chambers work on the same principles as bubble chambers, but are based on supersaturated vapor rather than superheated liquid. While bubble chambers were extensively used in the past, they have now mostly been supplanted by wire chambers and spark chambers. Historically, notable bubble chambers include the Big European Bubble Chamber (BEBC) and Gargamelle.

6.1   Function and use

The bubble chamber is similar to a cloud chamber, both in application and in basic principle. It is normally made by filling a large cylinder with a liquid heated to just below its boiling point. As particles enter the chamber, a piston suddenly decreases its pressure, and the liquid enters into a superheated, metastable phase. Charged particles create an ionization track, around which the liquid vaporizes, forming microscopic bubbles. Bubble density around a track is proportional to a particle's energy loss.

Bubbles grow in size as the chamber expands, until they are large enough to be seen or photographed. Several cameras are mounted around it, allowing a three-dimensional image of an event to be captured. Bubble chambers with resolutions down to a few μm have been operated.

The entire chamber is subject to a constant magnetic field, which causes charged particles to travel in helical paths whose radius is determined by their charge-to-mass ratios and their velocities. Since the magnitude of the charge of all known charged, long-lived subatomic particles is the same as that of an electron, their radius of curvature must be proportional to their momentum. Thus, by measuring their radius of curvature, their momentum can be determined.

Notable discoveries made by bubble chamber include the discovery of weak neutral currents at Gargamelle in 1973,[4] which established the soundness of the electroweak theory and led to the discovery of the W and Z bosons in 1983 (at the UA1 and UA2 experiments). Recently, bubble chambers have been used in research on WIMPs, at COUPP and PICASSO.[5][6]

6.2   Drawbacks

Although bubble chambers were very successful in the past, they are of only limited use in modern very-high-energy experiments, for a variety of reasons:

Fermilab's disused 15-foot bubble chamber

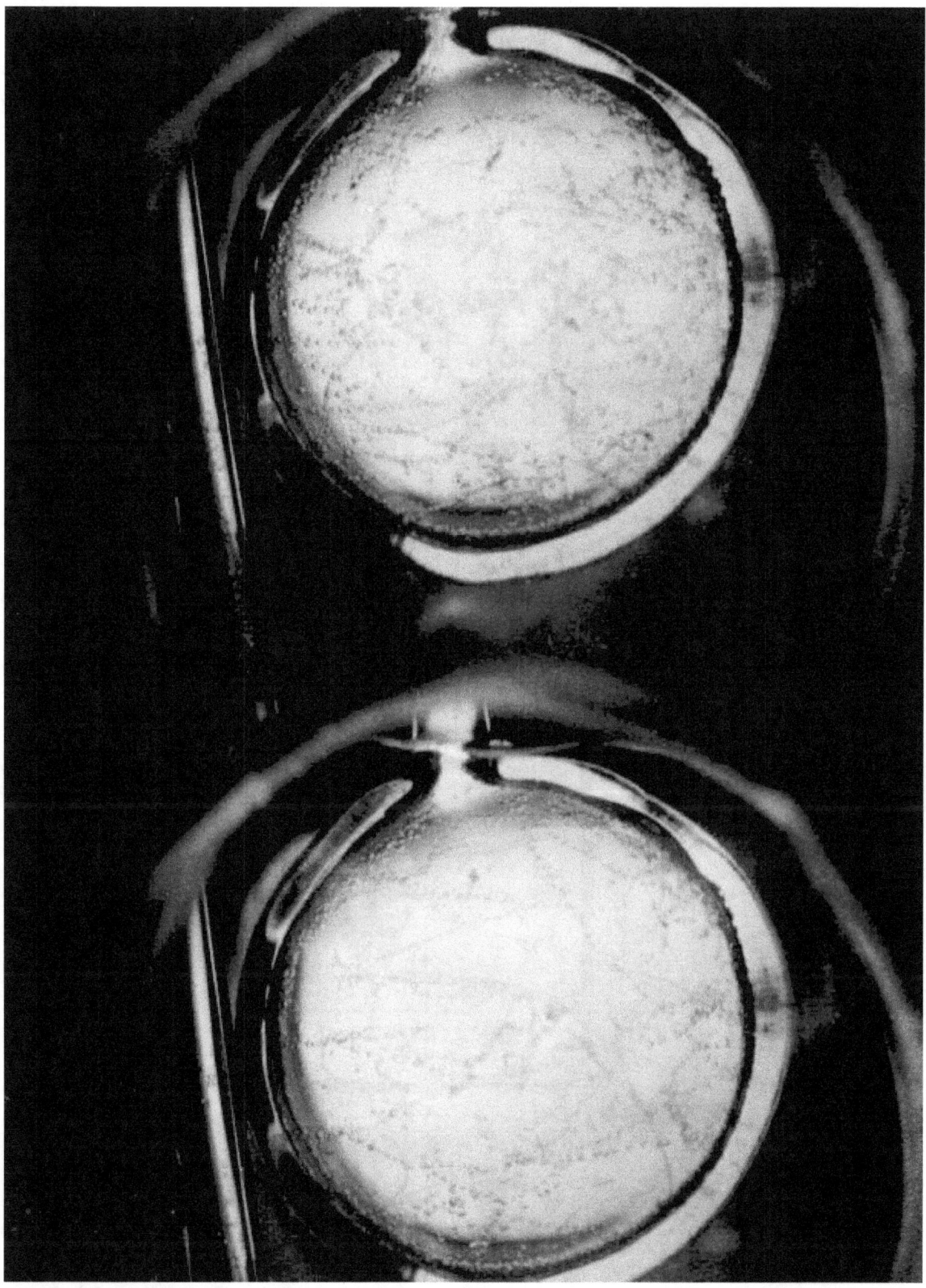

The first tracks observed in John Wood's 1.5-inch (~3.8 cm) liquid hydrogen bubble chamber, in 1954.

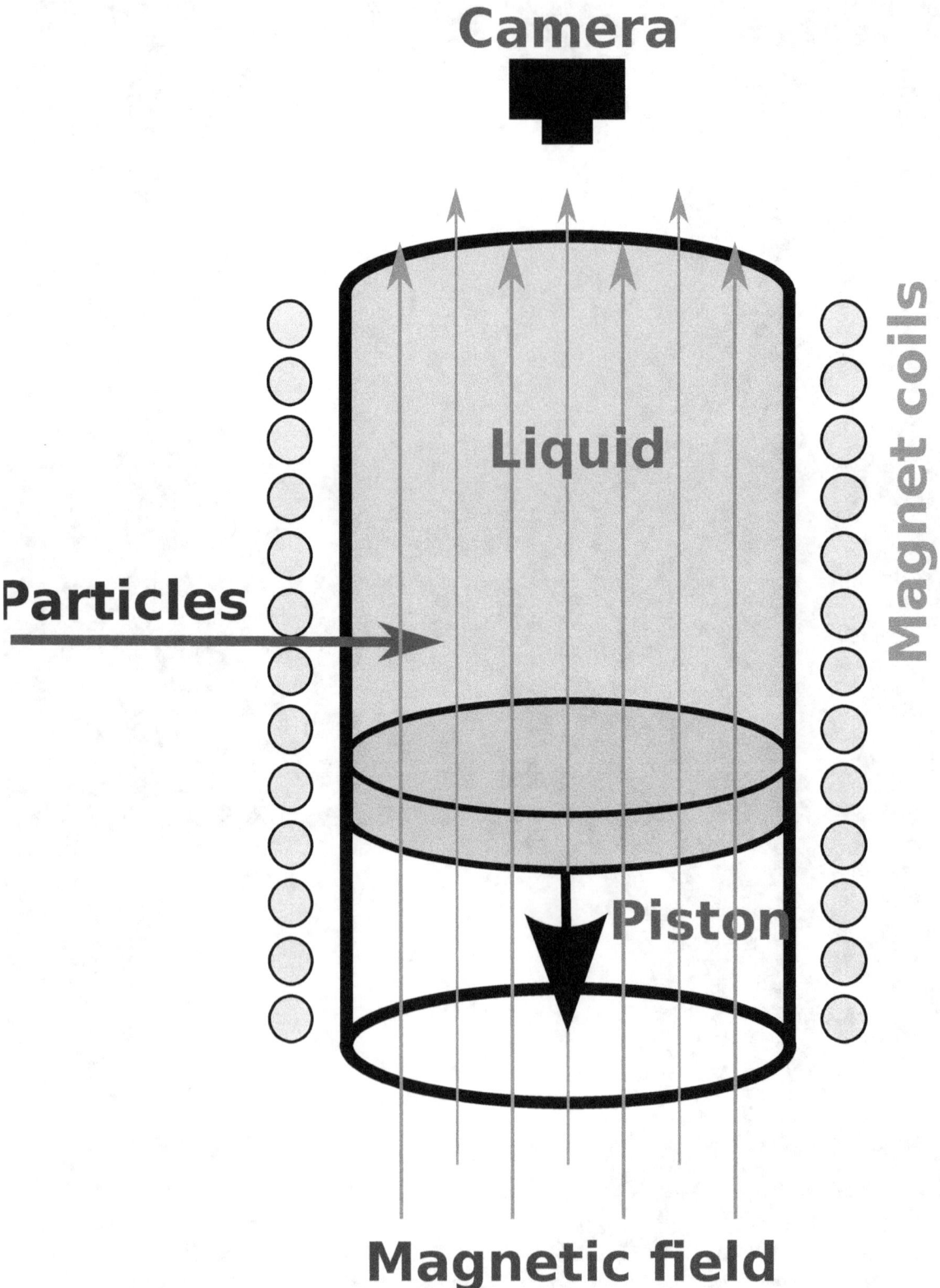

A bubble chamber

- The need for a photographic readout rather than three-dimensional electronic data makes it less convenient, especially in experiments which must be reset, repeated and analyzed many times.

- The superheated phase must be ready at the precise moment of collision, which complicates the detection of short-lived particles.

- Bubble chambers are neither large nor massive enough to analyze high-energy collisions, where all products should be contained inside the detector.

- The high-energy particles may have path radii too large to be accurately measured in a relatively small chamber, thereby hindering precise estimation of momentum.

Due to these issues, bubble chambers have largely been replaced by wire chambers, which allow particle energies to be measured at the same time. Another alternative technique is the spark chamber.

6.3 Notes

[1] Donald A. Glaser (1952). "Some Effects of Ionizing Radiation on the Formation of Bubbles in Liquids". *Physical Review* **87** (4): 665–665. Bibcode:1952PhRv...87..665G. doi:10.1103/PhysRev.87.665.

[2] "The Nobel Prize in Physics 1960". The Nobel Foundation. Retrieved 2009-10-03.

[3] Anne Pinckard (21 July 2006). "Front Seat to History: Summer Lecture Series Kicks Off – Invention and History of the Bubble Chamber". *Berkeley Lab View Archive*. Lawrence Berkeley National Laboratory. Retrieved 2009-10-03.

[4] "1973: Neutral currents are revealed". CERN. Retrieved 2009-10-03.

[5] "COUPP experiment – E961". COUPP. Retrieved 2009-10-03.

[6] "The PICASSO experiment". PICASSO. Retrieved 2009-10-03.

6.4 External links

- "A step-by-step tutorial on how to read bubble chamber pictures". CERN.

Chapter 7

Propagator

This article is about Quantum field theory. For plant propagation, see Plant propagation.

In quantum mechanics and quantum field theory, the **propagator** gives the probability amplitude for a particle to travel from one place to another in a given time, or to travel with a certain energy and momentum. In Feynman diagrams, which calculate the rate of collisions in quantum field theory, virtual particles contribute their propagator to the rate of the scattering event described by the diagram. They also can be viewed as the inverse of the wave operator appropriate to the particle, and are therefore often called *(causal) Green's functions* (called "*causal*" to distinguish it from the elliptic Laplacian Green's function).[1][2]

7.1 Non-relativistic propagators

In non-relativistic quantum mechanics the propagator gives the probability amplitude for a particle to travel from one spatial point at one time to another spatial point at a later time. It is the Green's function (fundamental solution) for the Schrödinger equation. This means that, if a system has Hamiltonian H, then the appropriate propagator is a function

$$G(x,t;x',t') = \frac{1}{i\hbar}\Theta(t-t')K(x,t;x',t')$$

satisfying

$$\left(i\hbar\frac{\partial}{\partial t} - H_x\right)G(x,t;x',t') = \delta(x-x')\delta(t-t')\,,$$

where Hx denotes the Hamiltonian written in terms of the x coordinates, $\delta(x)$ denotes the Dirac delta-function, $\Theta(x)$ is the Heaviside step function and $K(x, t\ ;x', t')$ is the kernel of the differential operator in question, often referred to as the propagator instead of G in this context, and henceforth in this article. This propagator can also be written as

$$K(x,t;x',t') = \left\langle x|\hat{U}(t,t')|x'\right\rangle,$$

where $\hat{U}(t, t')$ is the unitary time-evolution operator for the system taking states at time t to states at time t'.

The quantum mechanical propagator may also be found by using a path integral,

$$K(x,t;x',t') = \int \exp\left[\frac{i}{\hbar}\int_t^{t'} L(\dot{q},q,t)dt\right]D[q(t)]$$

where the boundary conditions of the path integral include $q(t) = x$, $q(t')=x'$. Here L denotes the Lagrangian of the system. The paths that are summed over move only forwards in time.

In non-relativistic quantum mechanics, the propagator lets you find the state of a system given an initial state and a time interval. The new state is given by the equation

$$\psi(x, t) = \int_{-\infty}^{\infty} \psi(x', t') K(x, t; x', t') dx'.$$

If $K(x,t;x',t')$ only depends on the difference $x - x'$, this is a convolution of the initial state and the propagator. This kernel is the kernel of integral transform.

7.1.1 Basic examples: propagator of free particle and harmonic oscillator

For a time-translationally invariant system, the propagator only depends on the time difference $t - t'$, so it may be rewritten as

$$K(x, t; x', t') = K(x, x'; t - t').$$

The propagator of a one-dimensional free particle, with the far-right expression obtained via saddle-point methods,[3] is then

Similarly, the propagator of a one-dimensional quantum harmonic oscillator is the Mehler kernel,[4][5]

The latter may be obtained from the previous free particle result upon making use of van Kortryk's SU(2) Lie-group identity,

$$\exp\left(-\frac{it}{\hbar}\left(\frac{1}{2m}\,\mathsf{p}^2 + \frac{1}{2}\,m\omega^2\mathsf{x}^2\right)\right)$$

$$= \exp\left(-\frac{im\omega}{2\hbar}\,\mathsf{x}^2 \tan\left(\frac{\omega t}{2}\right)\right) \exp\left(-\frac{i}{2m\omega\hbar}\,\mathsf{p}^2 \sin\left(\omega t\right)\right) \exp\left(-\frac{im\omega}{2\hbar}\,\mathsf{x}^2 \tan\left(\frac{\omega t}{2}\right)\right),$$

valid for operators x and p satisfying the Heisenberg relation $[\mathsf{x}, \mathsf{p}] = i\hbar$.

For the N-dimensional case, the propagator can be simply obtained by the product

$$K(\vec{x}, \vec{x}'; t) = \prod_{q=1}^{N} K(x_q, x'_q; t).$$

7.2 Relativistic propagators

In relativistic quantum mechanics and quantum field theory the propagators are Lorentz invariant. They give the amplitude for a particle to travel between two spacetime points.

7.2.1   Scalar propagator

In quantum field theory the theory of a free (non-interacting) scalar field is a useful and simple example which serves to illustrate the concepts needed for more complicated theories. It describes spin zero particles. There are a number of possible propagators for free scalar field theory. We now describe the most common ones.

7.2.2   Position space

The position space propagators are Green's functions for the Klein–Gordon equation. This means they are functions $G(x, y)$ which satisfy

$$(\Box_x + m^2)G(x,y) = -\delta(x - y)$$

where:

- x, y are two points in Minkowski spacetime.
- $\Box_x = \frac{\partial^2}{\partial t^2} - \nabla^2$ is the d'Alembertian operator acting on the x coordinates.
- $\delta(x - y)$ is the Dirac delta-function.

(As typical in relativistic quantum field theory calculations, we use units where the speed of light, c, and Planck's reduced constant, ℏ, are set to unity.)

We shall restrict attention to 4-dimensional Minkowski spacetime. We can perform a Fourier transform of the equation for the propagator, obtaining

$$\left(-p^2 + m^2\right) G(p) = -1.$$

This equation can be inverted in the sense of distributions noting that the equation xf(x)=1 has the solution, (see Sokhotski-Plemelj theorem)

$$f(x) = \frac{1}{x \pm i\varepsilon} = \frac{1}{x} \mp i\pi\delta(x),$$

with ε implying the limit to zero. Below, we discuss the right choice of the sign arising from causality requirements.

The solution is

$$G(x,y) = \frac{1}{(2\pi)^4} \int d^4p \, \frac{e^{-ip(x-y)}}{p^2 - m^2 \pm i\varepsilon}$$

where

$$p(x - y) := p_0(x^0 - y^0) - \vec{p} \cdot (\vec{x} - \vec{y})$$

is the 4-vector inner product.

The different choices for how to deform the integration contour in the above expression lead to different forms for the propagator. The choice of contour is usually phrased in terms of the p_0 integral.

The integrand then has two poles at

$$p_0 = \pm\sqrt{\vec{p}^2 + m^2}$$

so different choices of how to avoid these lead to different propagators.

Causal propagator

Retarded propagator

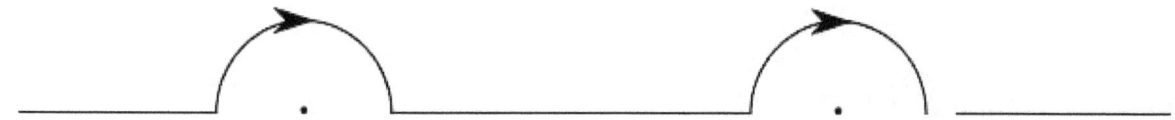

A contour going clockwise over both poles gives the **causal retarded propagator**. This is zero if x and y are spacelike or if $x^0 < y^0$ (i.e. if y is to the future of x).

This choice of contour is equivalent to calculating the limit,

$$G_{ret}(x,y) = \lim_{\epsilon \to 0} \frac{1}{(2\pi)^4} \int d^4p \, \frac{e^{-ip(x-y)}}{(p_0 + i\epsilon)^2 - \vec{p}^2 - m^2} = \begin{cases} \frac{1}{2\pi}\delta(\tau_{xy}^2) - \frac{mJ_1(m\tau_{xy})}{4\pi\tau_{xy}} & y \prec x \\ 0 & \text{otherwise} \end{cases}$$

Here

$$\tau_{xy} := \sqrt{(x^0 - y^0)^2 - (\vec{x} - \vec{y})^2}$$

is the proper time from x to y and J_1 is a Bessel function of the first kind. The expression $y \prec x$ means y causally precedes x which, for Minkowski spacetime, means

$$y^0 < x^0 \text{ and } \tau_{xy}^2 \geq 0 .$$

This expression can also be expressed in terms of the vacuum expectation value of the commutator of the free scalar field operator,

$$G_{ret}(x,y) = i\langle 0| \left[\Phi(x), \Phi(y)\right] |0\rangle \Theta(x^0 - y^0)$$

where

$$\Theta(x) := \begin{cases} 1 & x \geq 0 \\ 0 & x < 0 \end{cases}$$

is the Heaviside step function and

$$[\Phi(x), \Phi(y)] := \Phi(x)\Phi(y) - \Phi(y)\Phi(x)$$

is the commutator.

Advanced propagator

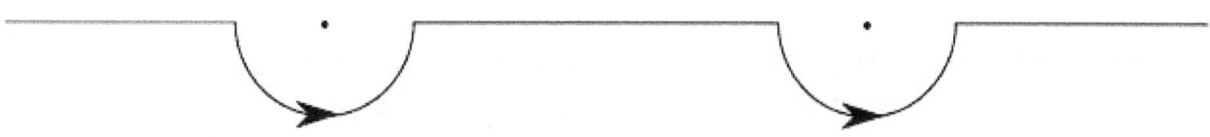

A contour going anti-clockwise under both poles gives the **causal advanced propagator**. This is zero if x and y are spacelike or if $x^0 > y^0$ (i.e. if y is to the past of x).

This choice of contour is equivalent to calculating the limit

$$G_{adv}(x,y) = \lim_{\epsilon \to 0} \frac{1}{(2\pi)^4} \int d^4p \, \frac{e^{-ip(x-y)}}{(p_0 - i\epsilon)^2 - \vec{p}^2 - m^2} = \begin{cases} -\frac{1}{2\pi}\delta(\tau_{xy}^2) + \frac{mJ_1(m\tau_{xy})}{4\pi\tau_{xy}} & x \prec y \\ 0 & \text{otherwise} \end{cases}$$

This expression can also be expressed in terms of the vacuum expectation value of the commutator of the free scalar field. In this case,

$$G_{adv}(x,y) = -i\langle 0|\,[\Phi(x),\Phi(y)]\,|0\rangle\Theta(y^0 - x^0)\,.$$

Feynman propagator

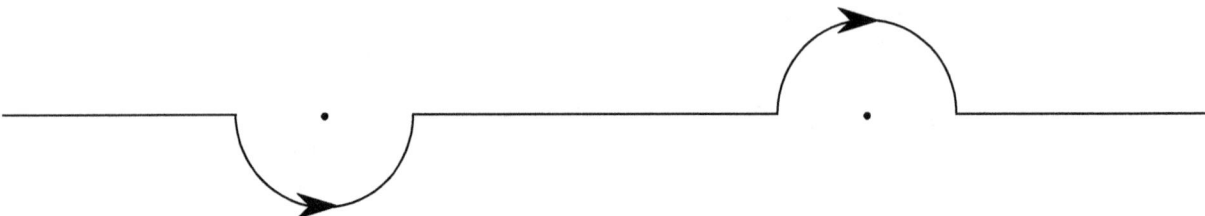

A contour going under the left pole and over the right pole gives the **Feynman propagator**.

This choice of contour is equivalent to calculating the limit (see Huang p. 30)

$$G_F(x,y) = \lim_{\epsilon \to 0} \frac{1}{(2\pi)^4} \int d^4p \, \frac{e^{-ip(x-y)}}{p^2 - m^2 + i\epsilon} = \begin{cases} -\frac{1}{4\pi}\delta(s) + \frac{m}{8\pi\sqrt{s}}H_1^{(2)}(m\sqrt{s}) & s \geq 0 \\ -\frac{im}{4\pi^2\sqrt{-s}}K_1(m\sqrt{-s}) & s < 0. \end{cases}$$

Here

$$s := (x^0 - y^0)^2 - (\vec{x} - \vec{y})^2,$$

where x and y are two points in Minkowski spacetime, and the dot in the exponent is a four-vector inner product. $H_1^{(2)}$ is a Hankel function and K_1 is a modified Bessel function.

This expression can be derived directly from the field theory as the vacuum expectation value of the *time-ordered product* of the free scalar field, that is, the product always taken such that the time ordering of the spacetime points is the same,

$$G_F(x-y) = -i\langle 0|T(\Phi(x)\Phi(y))|0\rangle = -i\,\langle 0|\,\left[\Theta(x^0 - y^0)\Phi(x)\Phi(y) + \Theta(y^0 - x^0)\Phi(y)\Phi(x)\right]|0\rangle\,.$$

This expression is Lorentz invariant, as long as the field operators commute with one another when the points x and y are separated by a spacelike interval.

The usual derivation is to insert a complete set of single-particle momentum states between the fields with Lorentz covariant normalization, then show that the Θ functions providing the causal time ordering may be obtained by a contour integral along the energy axis if the integrand is as above (hence the infinitesimal imaginary part, to move the pole off the real line).

The propagator may also be derived using the path integral formulation of quantum theory.

7.2.3   Momentum space propagator

The Fourier transform of the position space propagators can be thought of as propagators in momentum space. These take a much simpler form than the position space propagators.

They are often written with an explicit ε term although this is understood to be a reminder about which integration contour is appropriate (see above). This ε term is included to incorporate boundary conditions and causality (see below).

For a 4-momentum p the causal and Feynman propagators in momentum space are:

$$\tilde{G}_{ret}(p) = \frac{1}{(p_0 + i\varepsilon)^2 - \vec{p}^2 - m^2}$$

$$\tilde{G}_{adv}(p) = \frac{1}{(p_0 - i\varepsilon)^2 - \vec{p}^2 - m^2}$$

$$\tilde{G}_F(p) = \frac{1}{p^2 - m^2 + i\varepsilon}.$$

For purposes of Feynman diagram calculations it is usually convenient to write these with an additional overall factor of −i (conventions vary).

7.2.4 Faster than light?

The Feynman propagator has some properties that seem baffling at first. In particular, unlike the commutator, the propagator is *nonzero* outside of the light cone, though it falls off rapidly for spacelike intervals. Interpreted as an amplitude for particle motion, this translates to the virtual particle traveling faster than light. It is not immediately obvious how this can be reconciled with causality: can we use faster-than-light virtual particles to send faster-than-light messages?

The answer is no: while in classical mechanics the intervals along which particles and causal effects can travel are the same, this is no longer true in quantum field theory, where it is commutators that determine which operators can affect one another.

So what *does* the spacelike part of the propagator represent? In QFT the vacuum is an active participant, and particle numbers and field values are related by an uncertainty principle; field values are uncertain even for particle number *zero*. There is a nonzero probability amplitude to find a significant fluctuation in the vacuum value of the field $\Phi(x)$ if one measures it locally (or, to be more precise, if one measures an operator obtained by averaging the field over a small region). Furthermore, the dynamics of the fields tend to favor spatially correlated fluctuations to some extent. The nonzero time-ordered product for spacelike-separated fields then just measures the amplitude for a nonlocal correlation in these vacuum fluctuations, analogous to an EPR correlation. Indeed, the propagator is often called a *two-point correlation function* for the free field.

Since, by the postulates of quantum field theory, all observable operators commute with each other at spacelike separation, messages can no more be sent through these correlations than they can through any other EPR correlations; the correlations are in random variables.

In terms of virtual particles, the propagator at spacelike separation can be thought of as a means of calculating the amplitude for creating a virtual particle-antiparticle pair that eventually disappear into the vacuum, or for detecting a virtual pair emerging from the vacuum. In Feynman's language, such creation and annihilation processes are equivalent to a virtual particle wandering backward and forward through time, which can take it outside of the light cone. However, no causality violation is involved.

Explanation using Limits

This can be made clearer by writing the propagator in the following form for a massless photon:

$$G_F^\varepsilon(x, y) = \frac{\varepsilon}{(x - y)^2 + i\varepsilon^2}$$

This is the usual definition but normalised by a factor of ε. Then the rule is that one only takes the limit $\varepsilon \to 0$ at the end of a calculation. We easily see that

$$G_F^\varepsilon(x, y) = \tfrac{1}{\varepsilon} \text{ if } (x - y)^2 = 0$$

and

$$\lim_{\varepsilon \to 0} G_F^\varepsilon(x, y) = 0 \text{ if } (x - y)^2 \neq 0$$

Hence this means a single photon will always stay on the light cone. It is also shown that the total probability for a photon at any time must be normalised by the reciprocal of the following factor:

$$\lim_{\varepsilon \to 0} \int |G_F^\varepsilon(0, x)|^2 \, dx^3 = \lim_{\varepsilon \to 0} \int \frac{\varepsilon^2}{(\mathbf{x}^2 - t^2)^2 + \varepsilon^4} dx^3 = 2\pi^2 |t|$$

We see that the parts outside the light cone usually are zero in the limit and only are important in Feynman diagrams.

7.2.5   Propagators in Feynman diagrams

The most common use of the propagator is in calculating probability amplitudes for particle interactions using Feynman diagrams. These calculations are usually carried out in momentum space. In general, the amplitude gets a factor of the propagator for every *internal line*, that is, every line that does not represent an incoming or outgoing particle in the initial or final state. It will also get a factor proportional to, and similar in form to, an interaction term in the theory's Lagrangian for every internal vertex where lines meet. These prescriptions are known as *Feynman rules*.

Internal lines correspond to virtual particles. Since the propagator does not vanish for combinations of energy and momentum disallowed by the classical equations of motion, we say that the virtual particles are allowed to be off shell. In fact, since the propagator is obtained by inverting the wave equation, in general it will have singularities on shell.

The energy carried by the particle in the propagator can even be *negative*. This can be interpreted simply as the case in which, instead of a particle going one way, its antiparticle is going the *other* way, and therefore carrying an opposing flow of positive energy. The propagator encompasses both possibilities. It does mean that one has to be careful about minus signs for the case of fermions, whose propagators are not even functions in the energy and momentum (see below).

Virtual particles conserve energy and momentum. However, since they can be off shell, wherever the diagram contains a closed *loop*, the energies and momenta of the virtual particles participating in the loop will be partly unconstrained, since a change in a quantity for one particle in the loop can be balanced by an equal and opposite change in another. Therefore, every loop in a Feynman diagram requires an integral over a continuum of possible energies and momenta. In general, these integrals of products of propagators can diverge, a situation that must be handled by the process of renormalization.

7.2.6   Other theories

If the particle possesses spin then its propagator is in general somewhat more complicated, as it will involve the particle's spin or polarization indices. The momentum-space propagator used in Feynman diagrams for a Dirac field representing the electron in quantum electrodynamics has the form

$$\tilde{S}_F(p) = \frac{(\gamma^\mu p_\mu + m)}{p^2 - m^2 + i\varepsilon}$$

where the γ^μ are the gamma matrices appearing in the covariant formulation of the Dirac equation. It is sometimes written, using Feynman slash notation,

$$\tilde{S}_F(p) = \frac{1}{\gamma^\mu p_\mu - m + i\varepsilon} = \frac{1}{\not{p} - m + i\epsilon}$$

for short. In position space we have:

$$S_F(x - y) = \int \frac{d^4p}{(2\pi)^4} e^{-ip \cdot (x-y)} \frac{\gamma^\mu p_\mu + m}{p^2 - m^2 + i\epsilon} = \left(\frac{\gamma^\mu (x - y)_\mu}{|x - y|^5} + \frac{m}{|x - y|^3} \right) J_1(m|x - y|).$$

This is related to the Feynman propagator by

$$S_F(x - y) = (i\partial\!\!\!/ + m)G_F(x - y)$$

where $\partial\!\!\!/ := \gamma^\mu \partial_\mu$.

The propagator for a gauge boson in a gauge theory depends on the choice of convention to fix the gauge. For the gauge used by Feynman and Stueckelberg, the propagator for a photon is

$$\frac{-ig^{\mu\nu}}{p^2 + i\epsilon}.$$

The propagator for a massive vector field can be derived from the Stueckelberg Lagrangian. The general form with gauge parameter λ reads

$$\frac{g_{\mu\nu} - \frac{k_\mu k_\nu}{m^2}}{k^2 - m^2 + i\epsilon} + \frac{\frac{k_\mu k_\nu}{m^2}}{k^2 - \frac{m^2}{\lambda} + i\epsilon}.$$

With this general form one obtains the propagator in unitary gauge for $\lambda = 0$, the propagator in Feynman or 't Hooft gauge for $\lambda = 1$ and in Landau or Lorenz gauge for $\lambda = \infty$. There are also other notations where the gauge parameter is the inverse of λ. The name of the propagator however refers to its final form and not necessarily to the value of the gauge parameter.

Unitary gauge:

$$\frac{g_{\mu\nu} - \frac{k_\mu k_\nu}{m^2}}{k^2 - m^2 + i\epsilon}.$$

Feynman ('t Hooft) gauge:

$$\frac{g_{\mu\nu}}{k^2 - m^2 + i\epsilon}.$$

Landau (Lorenz) gauge:

$$\frac{g_{\mu\nu} - \frac{k_\mu k_\nu}{k^2}}{k^2 - m^2 + i\epsilon}.$$

7.3   Related singular functions

The scalar propagators are Green's functions for the Klein–Gordon equation. There are related singular functions which are important in quantum field theory. We follow the notation in Bjorken and Drell.[6] See also Bogolyubov and Shirkov (Appendix A). These function are most simply defined in terms of the vacuum expectation value of products of field operators.

7.3.1   Solutions to the Klein–Gordon equation

Pauli–Jordan function

The commutator of two scalar field operators defines the Pauli–Jordan function $\Delta(x-y)$ by[6]

$$\langle 0 | \left[\Phi(x), \Phi(y) \right] | 0 \rangle = i\Delta(x-y)$$

with

$$\Delta(x-y) = G_{adv}(x-y) - G_{ret}(x-y)$$

This satisfies $\Delta(x-y) = -\Delta(y-x)$ and is zero if $(x-y)^2 < 0$.

Positive and negative frequency parts (cut propagators)

We can define the positive and negative frequency parts of $\Delta(x-y)$, sometimes called cut propagators, in a relativistically invariant way.

This allows us to define the positive frequency part:

$$\Delta_+(x-y) = \langle 0 | \Phi(x)\Phi(y) | 0 \rangle$$

and the negative frequency part:

$$\Delta_-(x-y) = \langle 0 | \Phi(y)\Phi(x) | 0 \rangle$$

These satisfy[6]

$$i\Delta = \Delta_+ - \Delta_-$$

and

$$(\Box_x + m^2)\Delta_\pm(x-y) = 0.$$

Auxiliary function

The anti-commutator of two scalar field operators defines $\Delta_1(x-y)$ function by

$$\langle 0 | \left\{ \Phi(x), \Phi(y) \right\} | 0 \rangle = \Delta_1(x-y)$$

with

$$\Delta_1(x-y) = \Delta_+(x-y) + \Delta_-(x-y).$$

This satisfies $\Delta_1(x-y) = \Delta_1(y-x)$.

7.3.2 Green's functions for the Klein-Gordon equation

The retarded, advanced and Feynman propagators defined above are all Green's functions for the Klein-Gordon equation. They are related to the singular functions by[6]

$$G_{ret}(x - y) = -\Delta(x - y)\Theta(x_0 - y_0)$$

$$G_{adv}(x - y) = \Delta(x - y)\Theta(y_0 - x_0)$$

$$2G_F(x - y) = -i\Delta_1(x - y) + \epsilon(x_0 - y_0)\Delta(x - y)$$

where $\epsilon(x_0 - y_0) = 2\Theta(x_0 - y_0) - 1$.

7.4 References

[1] The mathematics of PDEs and the wave equation, pg 32. Michael P. Lamoureux, University of Calgary, Seismic Imaging Summer School, August 7–11, 2006, Calgary.

[2] Ch.: 9 Green's functions, pg 6. FOURIER ANALYSIS LECTURE COURSE: LECTURE 15.

[3] Saddle point approximation, planetmath.org

[4] E. U. Condon, "Immersion of the Fourier transform in a continuous group of functional transformations", *Proc. Nat. Acad. Sci. USA* **23**, (1937) 158–164. online

[5] Wolfgang Pauli, *Wave Mechanics: Volume 5 of Pauli Lectures on Physics* (Dover Books on Physics, 2000) ISBN 0486414620 , cf. Section 44.

[6] Bjorken and Drell, Appendix C

- Bjorken, J.D., Drell, S.D., *Relativistic Quantum Fields* (Appendix C.), New York: McGraw-Hill 1965, ISBN 0-07-005494-0.

- N. N. Bogoliubov, D. V. Shirkov, *Introduction to the theory of quantized fields*, Wiley-Interscience, ISBN 0-470-08613-0 (Especially pp. 136–156 and Appendix A)

- Edited by DeWitt, Cécile and DeWitt, Bryce, *Relativity, Groups and Topology*, section Dynamical Theory of Groups & Fields, (Blackie and Son Ltd, Glasgow), Especially p615-624, ISBN 0-444-86858-5

- Griffiths, David J., *Introduction to Elementary Particles*, New York: John Wiley & Sons, 1987. ISBN 0-471-60386-4

- Griffiths, David J., *Introduction to Quantum Mechanics*, Upper Saddle River: Prentice Hall, 2004. ISBN 0-131-11892-7

- Halliwell, J.J., Orwitz, M. *Sum-over-histories origin of the composition laws of relativistic quantum mechanics and quantum cosmology*, arXiv:gr-qc/9211004

- Huang, Kerson, *Quantum Field Theory: From Operators to Path Integrals* (New York: J. Wiley & Sons, 1998), ISBN 0-471-14120-8

- Itzykson, Claude, Zuber, Jean-Bernard *Quantum Field Theory*, New York: McGraw-Hill, 1980. ISBN 0-07-032071-3

- Pokorski, Stefan, *Gauge Field Theories*, Cambridge: Cambridge University Press, 1987. ISBN 0-521-36846-4 *(Has useful appendices of Feynman diagram rules, including propagators, in the back.)*

- Schulman, Larry S., *Techniques & Applications of Path Integration*, Jonh Wiley & Sons (New York-1981) ISBN 0-471-76450-7

7.5   External links

- Three Methods for Computing the Feynman Propagator

Chapter 8

Interaction picture

In quantum mechanics, the **interaction picture** (also known as the **Dirac picture**) is an intermediate representation between the Schrödinger picture and the Heisenberg picture. Whereas in the other two pictures either the state vector or the operators carry time dependence, in the interaction picture both carry part of the time dependence of observables.[1] The interaction picture is useful in dealing with changes to the wave functions and observable due to interactions. Most field theoretical calculations[2] use the interaction representation because they construct the solution to the many body Schrödinger equation as the solution to the free particle problem plus some unknown interaction parts.

Equations that include operators acting at different times, which hold in the interaction picture, don't necessarily hold in the Schrödinger or the Heisenberg picture. This is because time-dependent unitary transformations relate operators in one picture to the analogous operators in the others.

8.1 Definition

Operators and state vectors in the interaction picture are related by a change of basis (unitary transformation) to those same operators and state vectors in the Schrödinger picture.

To switch into the interaction picture, we divide the Schrödinger picture Hamiltonian into two parts,

Any possible choice of parts will yield a valid interaction picture; but in order for the interaction picture to be useful in simplifying the analysis of a problem, the parts will typically be chosen so that $H_{0,S}$ is well understood and exactly solvable, while $H_{1,S}$ contains some harder-to-analyze perturbation to this system.

If the Hamiltonian has *explicit time-dependence* (for example, if the quantum system interacts with an applied external electric field that varies in time), it will usually be advantageous to include the explicitly time-dependent terms with $H_{1,S}$, leaving $H_{0,S}$ time-independent. We proceed assuming that this is the case. If there *is* a context in which it makes sense to have $H_{0,S}$ be time-dependent, then one can proceed by replacing $e^{\pm i H_{0,S} t/\hbar}$ by the corresponding time-evolution operator in the definitions below.

8.1.1 State vectors

A state vector in the interaction picture is defined as[3]

where $|\psi_S(t)\rangle$ is the state vector in the Schrödinger picture.

8.1.2 Operators

An operator in the interaction picture is defined as

Note that $AS(t)$ will typically not depend on t, and can be rewritten as just AS. It only depends on t if the operator has "explicit time dependence", for example due to its dependence on an applied, external, time-varying electric field.

Hamiltonian operator

For the operator H_0 itself, the interaction picture and Schrödinger picture coincide,

$$H_{0,I}(t) = e^{iH_{0,S}t/\hbar} H_{0,S} e^{-iH_{0,S}t/\hbar} = H_{0,S}.$$

This is easily seen through the fact that operators commute with differentiable functions of themselves. This particular operator then can be called H_0 without ambiguity.

For the perturbation Hamiltonian H_1,I, however,

$$H_{1,I}(t) = e^{iH_{0,S}t/\hbar} H_{1,S} e^{-iH_{0,S}t/\hbar},$$

where the interaction picture perturbation Hamiltonian becomes a time-dependent Hamiltonian—unless $[H_1,S, H_0,S] = 0$.

It is possible to obtain the interaction picture for a time-dependent Hamiltonian $H_{0,S}(t)$ as well, but the exponentials need to be replaced by the unitary propagator for the evolution generated by $H_{0,S}(t)$, or more explicitly with a time-ordered exponential integral.

Density matrix

The density matrix can be shown to transform to the interaction picture in the same way as any other operator. In particular, let ϱI and ϱS be the density matrix in the interaction picture and the Schrödinger picture, respectively. If there is probability pn to be in the physical state $|\psi n⟩$, then

$$\rho_I(t) = \sum_n p_n(t)|\psi_{n,I}(t)\rangle\langle\psi_{n,I}(t)| = \sum_n p_n(t)e^{iH_{0,S}t/\hbar}|\psi_{n,S}(t)\rangle\langle\psi_{n,S}(t)|e^{-iH_{0,S}t/\hbar} = e^{iH_{0,S}t/\hbar}\rho_S(t)e^{-iH_{0,S}t/\hbar}.$$

8.2 Time-evolution equations in the interaction picture

8.2.1 Time-evolution of states

Transforming the Schrödinger equation into the interaction picture gives:

$$i\hbar\frac{d}{dt}|\psi_I(t)\rangle = H_{1,I}(t)|\psi_I(t)\rangle.$$

This equation is referred to as the **Schwinger–Tomonaga equation**.

8.2.2 Time-evolution of operators

If the operator *A*S is time independent (i.e., does not have "explicit time dependence"; see above), then the corresponding time evolution for *A*I(*t*) is given by

$$i\hbar \frac{d}{dt} A_I(t) = [A_I(t), H_0].$$

In the interaction picture the operators evolve in time like the operators in the Heisenberg picture with the Hamiltonian $H' = H_0$.

8.2.3 Time-evolution of the density matrix

Transforming the Schwinger–Tomonaga equation into the language of the density matrix (or equivalently, transforming the von Neumann equation into the interaction picture) gives:

$$i\hbar \frac{d}{dt} \rho_I(t) = [H_{1,I}(t), \rho_I(t)].$$

8.3 Use of interaction picture

The purpose of the interaction picture is to shunt all the time dependence due to H_0 onto the operators, thus allowing them to evolve freely, and leaving only H_1,I to control the time-evolution of the state vectors.

The interaction picture is convenient when considering the effect of a small interaction term, H_1,S, being added to the Hamiltonian of a solved system, H_0,S. By utilizing the interaction picture, one can use time-dependent perturbation theory to find the effect of H_1,I,[4]:355ff e.g., in the derivation of Fermi's golden rule,[4]:359–363 or the Dyson series,[4]:355–357 in quantum field theory: In 1947, Tomonaga and Schwinger appreciated that covariant perturbation theory could be formulated elegantly in the interaction picture, since field operators can evolve in time as free fields, even in the presence of interactions, now treated perturbatively in such a Dyson series.

8.4 References

[1] Albert Messiah (1966). *Quantum Mechanics*, North Holland, John Wiley & Sons. ISBN 0486409244 ; J. J. Sakurai (1994). *Modern Quantum Mechanics* (Addison-Wesley) ISBN 9780201539295 .

[2] J. W. Negele, H. Orland (1988), Quantum Many-particle Systems, ISBN 0738200522

[3] The Interaction Picture, lecture notes from New York University

[4] Sakukrai, J. J.; Napolitano, Jim (2010), *Modern Quantum Mechanics* (2nd ed.), Addison-Wesley, ISBN 978-0805382914

- L.D. Landau; E.M. Lifshitz (1977). *Quantum Mechanics: Non-Relativistic Theory*. Vol. 3 (3rd ed.). Pergamon Press. ISBN 978-0-08-020940-1. Online copy

- Townsend, John S. (2000). *A Modern Approach to Quantum Mechanics, 2nd ed.* Sausalito, California: University Science Books. ISBN 1-891389-13-0.

8.5   See also

- Bra–ket notation

- Schrödinger equation

- Haag's theorem

Chapter 9

Schrödinger picture

In physics, the **Schrödinger picture** (also called the **Schrödinger representation**[1]) is a formulation of quantum mechanics in which the state vectors evolve in time, but the operators (observables and others) are constant with respect to time.[2][3] This differs from the Heisenberg picture which keeps the states constant while the observables evolve in time, and from the interaction picture in which both the states and the observables evolve in time. The Schrödinger and Heisenberg pictures are related as active and passive transformations and commutation relations between operators are preserved in the passage between the two pictures.

In the Schrödinger picture, the state of a system evolves with time. The evolution for a closed quantum system is brought about by a unitary operator, the time evolution operator. For time evolution from a state vector $|\psi(t_0)\rangle$ at time t_0 to a state vector $|\psi(t)\rangle$ at time t, the time-evolution operator is commonly written $U(t, t_0)$, and one has

$$|\psi(t)\rangle = U(t, t_0)|\psi(t_0)\rangle.$$

In the case where the Hamiltonian of the system does not vary with time, the time-evolution operator has the form

$$U(t, t_0) = e^{-iH(t-t_0)/\hbar},$$

where the exponent is evaluated via its Taylor series.

The Schrödinger picture is useful when dealing with a time-independent Hamiltonian H; that is, $\partial_t H = 0$.

9.1   Background

In elementary quantum mechanics, the state of a quantum-mechanical system is represented by a complex-valued wavefunction $\psi(x,t)$. More abstractly, the state may be represented as a state vector, or *ket*, $|\psi\rangle$.
This ket is an element of a *Hilbert space*, a vector space containing all possible states of the system. A quantum-mechanical operator is a function which takes a ket $|\psi\rangle$ and returns some other ket $|\psi'\rangle$.

The differences between the Schrödinger and Heisenberg pictures of quantum mechanics revolve around how to deal with systems that evolve in time: the time-dependent nature of the system *must* be carried by some combination of the state vectors and the operators. For example, a quantum harmonic oscillator may be in a state $|\psi\rangle$ for which the expectation value of the momentum, $\langle\psi|\hat{p}|\psi\rangle$, oscillates sinusoidally in time. One can then ask whether this sinusoidal oscillation should be reflected in the state vector $|\psi\rangle$, the momentum operator $\hat{p}$, or both. All three of these choices are valid; the first gives the Schrödinger picture, the second the Heisenberg picture, and the third the interaction picture.

Erwin Schrödinger (1887 – 1961)
Schrödinger shared the 1933 Nobel Prize in physics together with Paul Dirac for his contributions to quantum mechanics.

9.2   The time evolution operator

9.2.1 Definition

The time-evolution operator $U(t, t_0)$ is defined as the operator which acts on the ket at time t_0 to produce the ket at some other time t:

$$|\psi(t)\rangle = U(t, t_0)|\psi(t_0)\rangle.$$

For bras, we instead have

$$\langle\psi(t)| = \langle\psi(t_0)|U^\dagger(t, t_0).$$

9.2.2 Properties

Unitarity

The time evolution operator must be unitary. This is because we demand that the norm of the state ket must not change with time. That is,

$$\langle\psi(t)|\psi(t)\rangle = \langle\psi(t_0)|U^\dagger(t, t_0)U(t, t_0)|\psi(t_0)\rangle = \langle\psi(t_0)|\psi(t_0)\rangle.$$

Therefore,

$$U^\dagger(t, t_0)U(t, t_0) = I.$$

Identity

When $t = t_0$, U is the identity operator, since

$$|\psi(t_0)\rangle = U(t_0, t_0)|\psi(t_0)\rangle.$$

Closure

Time evolution from t_0 to t may be viewed as a two-step time evolution, first from t_0 to an intermediate time t_1, and then from t_1 to the final time t. Therefore,

$$U(t, t_0) = U(t, t_1)U(t_1, t_0).$$

9.2.3 Differential equation for time evolution operator

We drop the t_0 index in the time evolution operator with the convention that $t_0 = 0$ and write it as $U(t)$. The Schrödinger equation is

$$i\hbar\frac{\partial}{\partial t}|\psi(t)\rangle = H|\psi(t)\rangle,$$

where H is the Hamiltonian. Now using the time-evolution operator U to write $|\psi(t)\rangle = U(t)|\psi(0)\rangle$, we have

$$i\hbar\frac{\partial}{\partial t}U(t)|\psi(0)\rangle = HU(t)|\psi(0)\rangle.$$

Since $|\psi(0)\rangle$ is a constant ket (the state ket at $t = 0$), and since the above equation is true for any constant ket in the Hilbert space, the time evolution operator must obey the equation

$$i\hbar\frac{\partial}{\partial t}U(t) = HU(t).$$

If the Hamiltonian is independent of time, the solution to the above equation is[note 1]

$$U(t) = e^{-iHt/\hbar}.$$

Since H is an operator, this exponential expression is to be evaluated via its Taylor series:

$$e^{-iHt/\hbar} = 1 - \frac{iHt}{\hbar} - \frac{1}{2}\left(\frac{Ht}{\hbar}\right)^2 + \cdots.$$

Therefore,

$$|\psi(t)\rangle = e^{-iHt/\hbar}|\psi(0)\rangle.$$

Note that $|\psi(0)\rangle$ is an arbitrary ket. However, if the initial ket is an eigenstate of the Hamiltonian, with eigenvalue E, we get:

$$|\psi(t)\rangle = e^{-iEt/\hbar}|\psi(0)\rangle.$$

Thus we see that the eigenstates of the Hamiltonian are *stationary states*: they only pick up an overall phase factor as they evolve with time.

If the Hamiltonian is dependent on time, but the Hamiltonians at different times commute, then the time evolution operator can be written as

$$U(t) = \exp\left(-\frac{i}{\hbar}\int_0^t H(t')\,dt'\right),$$

If the Hamiltonian is dependent on time, but the Hamiltonians at different times do not commute, then the time evolution operator can be written as

$$U(t) = \mathrm{T}\exp\left(-\frac{i}{\hbar}\int_0^t H(t')\,dt'\right),$$

where T is time-ordering operator, which is sometimes known as the Dyson series, after F.J.Dyson.

The alternative to the Schrödinger picture is to switch to a rotating reference frame, which is itself being rotated by the propagator. Since the undulatory rotation is now being assumed by the reference frame itself, an undisturbed state function appears to be truly static. This is the Heisenberg picture.

9.3 See also

- Hamilton–Jacobi equation

- Interaction picture

- Heisenberg picture

9.4 Notes

[1] Here we use the fact that at $t = 0$, $U(t)$ must reduce to the identity operator.

[1] "Schrödinger representation". Encyclopedia of Mathematics. Retrieved 3 September 2013.

[2] Parker, C.B. (1994). *McGraw Hill Encyclopaedia of Physics* (2nd ed.). McGraw Hill. pp. 786, 1261. ISBN 0-07-051400-3.

[3] Y. Peleg, R. Pnini, E. Zaarur, E. Hecht (2010). *Quantum mechanics*. Schuam's outline series (2nd ed.). McGraw Hill. p. 70. ISBN 9-780071-623582.

9.5 Further reading

- *Principles of Quantum Mechanics* by R. Shankar, Plenum Press.

- *Modern Quantum mechanics* by J.J. Sakurai.

Chapter 10

Heisenberg picture

In physics, the **Heisenberg picture** (also called the **Heisenberg representation**[1]) is a formulation (largely due to Werner Heisenberg in 1925) of quantum mechanics in which the operators (observables and others) incorporate a dependency on time, but the state vectors are time-independent, an arbitrary fixed basis rigidly underlying the theory.

It stands in contrast to the Schrödinger picture in which the operators are constant, instead, and the states evolve in time. The two pictures only differ by a basis change with respect to time-dependency, which corresponds to the difference between active and passive transformations. The Heisenberg picture is the formulation of matrix mechanics in an arbitrary basis, in which the Hamiltonian is not necessarily diagonal.

It further serves to define a third, hybrid, picture, the Interaction picture.

10.1   Mathematical details

In the Heisenberg picture of quantum mechanics the state vectors, $|\psi(t)\rangle$, do not change with time, while observables A satisfy

where H is the Hamiltonian and $[\bullet,\bullet]$ denotes the commutator of two operators (in this case H and A). Taking expectation values automatically yields the Ehrenfest theorem, featured in the correspondence principle.

By the Stone–von Neumann theorem, the Heisenberg picture and the Schrödinger picture are unitarily equivalent, just a basis change in Hilbert space. In some sense, the Heisenberg picture is more natural and convenient than the equivalent Schrödinger picture, especially for relativistic theories. Lorentz invariance is manifest in the Heisenberg picture, since the state vectors do not single out the time or space.

This approach also has a more direct similarity to classical physics: by simply replacing the commutator above by the Poisson bracket, the **Heisenberg equation** reduces to an equation in Hamiltonian mechanics.

10.2   Derivation of Heisenberg's equation

For didactical reasons, the Heisenberg picture is introduced here from the subsequent, but more familiar, Schrödinger picture. The expectation value of an observable A, which is a Hermitian linear operator, for a given Schrödinger state $|\psi(t)\rangle$, is given by

$$\langle A \rangle_t = \langle \psi(t)|A|\psi(t)\rangle.$$

In the Schrödinger picture, the state $|\psi(t)\rangle$ at time t is related to the state $|\psi(0)\rangle$ at time 0 by a unitary time-evolution operator, $U(t)$,

$$|\psi(t)\rangle = U(t)|\psi(0)\rangle.$$

If the Hamiltonian does not vary with time, then the time-evolution operator can be written as

$$U(t) = e^{-iHt/\hbar},$$

where H is the Hamiltonian and $\hbar$ is the reduced Planck constant. Therefore,

$$\langle A \rangle_t = \langle \psi(0) | e^{iHt/\hbar} A e^{-iHt/\hbar} | \psi(0) \rangle.$$

Peg all state vectors to a rigid basis of $|\psi(0)\rangle$ then, and define

$$A(t) := e^{iHt/\hbar} A e^{-iHt/\hbar}.$$

It now follows that

$$
\begin{aligned}
\frac{d}{dt} A(t) &= \frac{i}{\hbar} H e^{iHt/\hbar} A e^{-iHt/\hbar} + e^{iHt/\hbar} \left(\frac{\partial A}{\partial t} \right) e^{-iHt/\hbar} + \frac{i}{\hbar} e^{iHt/\hbar} A \cdot (-H) e^{-iHt/\hbar} \\
&= \frac{i}{\hbar} e^{iHt/\hbar} \left(HA - AH \right) e^{-iHt/\hbar} + e^{iHt/\hbar} \left(\frac{\partial A}{\partial t} \right) e^{-iHt/\hbar} \\
&= \frac{i}{\hbar} \left(HA(t) - A(t)H \right) + e^{iHt/\hbar} \left(\frac{\partial A}{\partial t} \right) e^{-iHt/\hbar}.
\end{aligned}
$$

Differentiation was according to the product rule, while $\partial A/\partial t$ is the time derivative of the initial A, not the $A(t)$ operator defined. The last equation holds since $\exp(-iHt/\hbar)$ commutes with H.

Thus

$$\frac{d}{dt} A(t) = \frac{i}{\hbar} [H, A(t)] + e^{iHt/\hbar} \left(\frac{\partial A}{\partial t} \right) e^{-iHt/\hbar},$$

and hence emerges the above Heisenberg equation of motion, since the convective functional dependence on $x(0)$ and $p(0)$ converts to the *same* dependence on $x(t)$, $p(t)$, so that the last term converts to $\partial A(t)/\partial t$. $[X, Y]$ is the commutator of two operators and is defined as $[X, Y] := XY - YX$.

The equation is solved by the $A(t)$ defined above, as evident by use of the standard operator identity,

$$e^B A e^{-B} = A + [B, A] + \frac{1}{2!}[B, [B, A]] + \frac{1}{3!}[B, [B, [B, A]]] + \cdots.$$

which implies

$$A(t) = A + \frac{it}{\hbar}[H, A] - \frac{t^2}{2!\hbar^2}[H, [H, A]] - \frac{it^3}{3!\hbar^3}[H, [H, [H, A]]] + \ldots$$

This relation also holds for classical mechanics, the classical limit of the above, given the correspondence between Poisson brackets and commutators,

$$[A, H] \leftrightarrow i\hbar\{A, H\}$$

In classical mechanics, for an A with no explicit time dependence,

$$\{A, H\} = \frac{d}{dt}A ,$$

so, again, the expression for $A(t)$ is the Taylor expansion around $t = 0$.

In effect, the arbitrary rigid Hilbert space basis $|\psi(0)\rangle$ has receded from view, and is only considered at the very last step of taking specific expectation values or matrix elements of observables.

10.3   Commutator relations

Commutator relations may look different than in the Schrödinger picture, because of the time dependence of operators. For example, consider the operators $x(t_1)$, $x(t_2)$, $p(t_1)$ and $p(t_2)$. The time evolution of those operators depends on the Hamiltonian of the system. Considering the one-dimensional harmonic oscillator,

$$H = \frac{p^2}{2m} + \frac{m\omega^2 x^2}{2}$$

the evolution of the position and momentum operators is given by:

$$\frac{d}{dt}x(t) = -\frac{i}{\hbar}[H, x(t)] = \frac{p}{m}$$

$$\frac{d}{dt}p(t) = \frac{i}{\hbar}[H, p(t)] = -m\omega^2 x$$

Differentiating both equations once more and solving for them with proper initial conditions,

$$\dot{p}(0) = -m\omega^2 x_0,$$

$$\dot{x}(0) = \frac{p_0}{m},$$

leads to

$$x(t) = x_0 \cos(\omega t) + \frac{p_0}{\omega m} \sin(\omega t)$$

$$p(t) = p_0 \cos(\omega t) - m\omega x_0 \sin(\omega t)$$

Direct computation yields the more general commutator relations,

$$[x(t_1), x(t_2)] = \frac{i\hbar}{m\omega} \sin(\omega t_2 - \omega t_1)$$

$$[p(t_1), p(t_2)] = i\hbar m\omega \sin(\omega t_2 - \omega t_1)$$

$$[x(t_1), p(t_2)] = i\hbar \cos(\omega t_2 - \omega t_1)$$

For $t_1 = t_2$, one simply recovers the standard canonical commutation relations valid in all pictures.

10.4 Summary comparison of evolution in all pictures

10.5 See also

- Bra–ket notation
- Interaction picture
- Schrödinger picture
- Heisenberg-Langevin equation

10.6 References

[1] "Heisenberg representation". Encyclopedia of Mathematics. Retrieved 3 September 2013.

- Cohen-Tannoudji, Claude; Bernard Diu; Frank Laloe (1977). *Quantum Mechanics (Volume One)*. Paris: Wiley. pp. 312–314. ISBN 0-471-16433-X.

- Albert Messiah, 1966. *Quantum Mechanics* (Vol. I), English translation from French by G. M. Temmer. North Holland, John Wiley & Sons.

10.7 External links

- Pedagogic Aides to Quantum Field Theory Click on the link for Chap. 2 to find an extensive, simplified introduction to the Heisenberg picture.

Werner Heisenberg

10.8 Text and image sources, contributors, and licenses

10.8.1 Text

- **Feynman diagram** *Source:* https://en.wikipedia.org/wiki/Feynman_diagram?oldid=696413886 *Contributors:* CYD, Bryan Derksen, XJaM, Nate Silva, Shii, Stevertigo, Edward, Michael Hardy, Looxix~enwiki, Ahoerstemeier, Ryan Cable, Nikai, Kaihsu, Charles Matthews, Ww, Doradus, Furrykef, Phys, Omegatron, Bevo, Bhiggs, BenRG, Peak, Rasmus Faber, Intangir, SC, Wikibot, Anthony, Tobias Bergemann, Danenberg, Ancheta Wis, Giftlite, Sj, Harp, Alison, Remy B, Vivektewary, Sigfpe, Beland, Pmanderson, Icairns, Sam Hocevar, Robin klein, AlexChurchill, Urvabara, Jkl, Pyrop, Guanabot, Pjacobi, Vsmith, Rspeer, Mal~enwiki, Bender235, Ben Standeven, Robert P. O'Shea, AnyFile, Jjk, Matt McIrvin, Scentoni, Tritium6, Mdd, Neonumbers, Rgclegg, Ferrierd, Mac Davis, Tony Sidaway, RJFJR, Drat, Dominic, Markko, Linas, LoopZilla, Daira Hopwood, Ketiltrout, Koavf, Zbxgscqf, Commander, Strait, Miserlou, Ligulem, RE, FlaBot, Moskvax, RobertG, Gnostic804, Acyso, DoomBringer, Chobot, YurikBot, JabberWok, Gaius Cornelius, Rodier, Anomalocaris, SEWilcoBot, Welsh, SCZenz, Ragesoss, Nubby, Larsobrien, Ejl, DRosenbach, Dna-webmaster, Tomj, Caco de vidro, GrinBot~enwiki, KasugaHuang, SmackBot, Tom Lougheed, GaeusOctavius, Chris the speller, Jjalexand, JustThisGuy, Pieter Kuiper, Complexica, Colonies Chris, Salmar, Wikipedia brown, Xiner, Huon, Tesseran, DJIndica, Eliyak, SilverStar, Xiphoris, JanBielawski, Bitwise, Beefyt, Paul venter, Joseph Solis in Australia, Wikifarzin, Patrickwooldridge, 8754865, CmdrObot, Van helsing, Jsmaye, Joelholdsworth, Myasuda, Cydebot, Xxanthippe, Michael C Price, Thijs!bot, Epbr123, Barticus88, Headbomb, Davidhorman, Shlomi Hillel, Leevclarke, Sluzzelin, AniRaptor2001, CosineKitty, Jameskeates, Bakken, SHCarter, Swpb, Warchef, Connor Behan, R'n'B, Choihei, Sefog, Chiswick Chap, Fylwind, Joshmt, Vyn, Brvman, Sheliak, Mulanhua, Quilbert, Rei-bot, Kevin Steinhardt, Mbusux, Richwil, Ptrslv72, Drschawrz, SieBot, Likebox, Taemyr, Staylor71, Martarius, WurmWoode, ChandlerMapBot, Sjdunn9, DragonBot, Chutsu, GlasGhost, DumZiBoT, TimothyRias, Mchaddock, PSimeon, SilvonenBot, Truthnlove, Out of Phase User, Metsavend, Download, Chamal N, Favonian, ChenzwBot, Barak Sh, Mikkim64, AgadaUrbanit, Conroy23, Alfie66, Legobot, Luckas-bot, Yobot, Amirobot, AnomieBOT, Piano non troppo, Xqbot, Srich32977, PhysicsR, Noamz, Seeleschneider, Createangelos, Kismalac, Ysyoon, Jonesey95, Nurefsan, Fizzotter, Meier99, Hickorybark, Earthandmoon, Aaivazis, EmausBot, John of Reading, Bookalign, WikitanvirBot, Bornerdogge, Maschen, Zueignung, Xanchester, Anagogist, Antiqueight, Smack the donkey, Pfeiferwalter, OCCullens, Bakkedal, ChrisGualtieri, Equatorbit, Mrmagikpants, Jamesx12345, Pjpeters, Mark viking, MutluMan, Someone not using his real name, UltraBird, Airwoz, Monkbot, Quogle, Cheweblaze, FivePillarPurist, Chemistry1111 and Anonymous: 127

- **Antiparticle** *Source:* https://en.wikipedia.org/wiki/Antiparticle?oldid=693880154 *Contributors:* AxelBoldt, CYD, Mav, Bryan Derksen, Andre Engels, Josh Grosse, Stevertigo, Mrwojo, Patrick, RTC, Paddu, CesarB, Nikai, Nikola Smolenski, Charles Matthews, The Anomebot, Wik, Omegatron, Bevo, Altenmann, Merovingian, Intangir, Wikibot, Martinwguy, Giftlite, Bogdanb, Harp, BenFrantzDale, Herbee, Spencer195, Fleminra, Jason Quinn, Zeimusu, Mako098765, Karol Langner, Mike Rosoft, Helohe, Rich Farmbrough, Guanabot, Pjacobi, Guanabot2, Mr. Billion, Joanjoc~enwiki, Kghose, Cmdrjameson, Giraffedata, Matt McIrvin, HasharBot~enwiki, Pediddle, Deror avi, Woohookitty, Mindmatrix, Wdyoung, GregorB, SeventyThree, Justin Ormont, Palica, Marudubshinki, Tevatron~enwiki, Rjwilmsi, Ae77, MZMcBride, KaiMartin, FlaBot, Krackpipe, Commander Nemet, Roboto de Ajvol, YurikBot, Borgx, Bambaiah, Zhaladshar, Spike Wilbury, Bota47, Terbospeed, Mkossick, Tim314, 小为 robot, SmackBot, FocalPoint, Alsandro, Srnec, Dauto, Octahedron80, Drphilharmonic, Marcus Brute, Vinaiwbot~enwiki, Jake-helliwell, Grumpyyoungman01, Newone, Mellery, Van helsing, Tim1988, Myasuda, Gogo Dodo, Goldencako, Thijs!bot, Headbomb, Tyco.skinner, JAnDbot, Steveprutz, Ferritecore, Jpod2, Singularity, Dbiel, TomasBat, Eternalmatt, Joshmt, DorganBot, Cuckooman4, VolkovBot, TXiKiBoT, Red Act, Anonymous Dissident, AlleborgoBot, SieBot, Likebox, RadicalOne, Flyer22 Reborn, KoenDelaere, Thomega, RW Marloe, BrightRoundCircle, Davidmosen, Jacob.jose, Anyeverybody, ClueBot, Diagramma Della Verita, Alexbot, Eeekster, Rishi.bedi, SilvonenBot, NellieBly, Lilaspastia, SkyLined, AkhtaBot, CarsracBot, Lightbot, Legobot, Luckas-bot, Yobot, Planlips, Csmallw, AnomieBOT, Citation bot, Vuerqex, ArthurBot, Xqbot, Omnipaedista, RibotBOT, Muhwang, EmausBot, John of Reading, L Kensington, Benazhack, ClueBot NG, Geekingreen, Mesoderm, Bibcode Bot, BG19bot, B wik, Mark Arsten, Rm1271, Penguinstorm300, Robotsheepboy, YFdyh-bot, 77Mike77, संजीव कुमार, Dert567, Monkbot, Dhm4444, Iapec, Nazo!nin, KasparBot and Anonymous: 102

- **List of Feynman diagrams** *Source:* https://en.wikipedia.org/wiki/List_of_Feynman_diagrams?oldid=595495523 *Contributors:* Welsh, SmackBot, Headbomb, Yobot and Anonymous: 1

- **Angular momentum diagrams (quantum mechanics)** *Source:* https://en.wikipedia.org/wiki/Angular_momentum_diagrams_(quantum_mechanics)?oldid=635773989 *Contributors:* Headbomb, David Eppstein, Mild Bill Hiccup, Omnipaedista, Maschen and Joeinwiki

- **Minkowski diagram***Source:*https://en.wikipedia.org/wiki/Minkowski_diagram?oldid=695167465*Contributors:*Patrick,SebastianHelm, Charles Matthews,Robbot,Aetheling,Wolfgangbeyer,Lupin,Ablewisuk,Rgdboer,Teorth,Apyule,Cpcjr,Burn,Bart133,DonQuixote,Dmitry Brant, PoccilScript,Mo-Al,Mathbot,DVdm,Siddhant,KSchutte,Escuerdo,Krea,JocK,Roger wilco,That Guy,From That Show!,Crystallina,SmackBot, Andyflux,Colonies Chris,OrphanBot,Serenity-Fr,JBel,Rook wave,Josh-Levin@ieee.org,Cronholm144,Dr Greg,CmdrObot,Myasuda,D.H, Openlander,Ais523,JAnDbot,Swpb,BeŽet,Belovedfreak,Biglovinb,HiraV,TXiKiBoT,Someguy1221,Pennstatephil,Sai2020,Jean-Christophe BENOIST,Juanmantoya,Smsarmad,Myrikhan,PlantTrees,Wwheaton,VQuakr,Namreh ekim,Anticipation ofa New Lover's Arrival, The,Addbot,D1d4,84user,Yobot,Bobw52,NOrbeck,FrescoBot,Paine Ellsworth,Duschi,Dewritech,Quondum,Frietjes,Bibcode Bot,Acmedogs, BattyBot,CarrieVS,DmVdx,Frinthruit,7Sidz,Myrral,Loraof,Tetra quark,Ari Ljd and Anonymous:42

- **Bubble chamber** *Source:* https://en.wikipedia.org/wiki/Bubble_chamber?oldid=669666497 *Contributors:* Bryan Derksen, XJaM, Michael Hardy, Ellywa, Aarchiba, Kevin Baas, Jiang, Stone, The Anomebot, Finlay McWalter, Sanders muc, Rholton, Harp, DragonflySixtyseven, Deglr6328, Aniboy2000, Jkl, .:Ajvol:., Jag123, Physicistjedi, HasharBot~enwiki, Matt Yohe, Keenan Pepper, Lectonar, Dirac1933, Goudzovski, YurikBot, Akamad, Bisqwit, Spike Wilbury, Tetracube, Scriber~enwiki, GrinBot~enwiki, Yamaguchi先生, Gilliam, Acdx, Mion, Vinaiwbot~enwiki, Andrei Stroe, Noah Salzman, Spiel496, Nfutvol, Jsmaye, PseudoNym, Thijs!bot, Headbomb, Sobreira, Electron9, Stannered, Martin62, Flavio Ribeiro, CommonsDelinker, Aqwis, DorganBot, Idioma-bot, VolkovBot, Philip Trueman, A4bot, PokeYourHeadOff, Suwatest, Fratrep, WikiBotas, ClueBot, Hostile Amish, DumZiBoT, Addbot, DOI bot, LaaknorBot, Lightbot, Zorrobot, Luckas-bot, Götz, Xqbot, RedBot, EmausBot, Torturella, Hhhippo, ClueBot NG, Bibcode Bot, Penguinstorm300, Trackteur, KasparBot and Anonymous: 35

- **Propagator** *Source:* https://en.wikipedia.org/wiki/Propagator?oldid=696771636 *Contributors:* Michael Hardy, Cyp, Jitse Niesen, Chuunen Baka, Giftlite, MathKnight, Lumidek, TobinFricke, Chris Howard, Pjacobi, MuDavid, Bjoern~enwiki, Smalljim, Matt McIrvin, PopUpPirate, Egg, Torquil~enwiki, SMC, R.e.b., Lionelbrits, Itinerant1, Commander Nemet, Welsh, Eudoxie, Kymara, Pretzels, Summentier, Lambiam,

SilkTork, Mhohmann, Lugnuts, Thijs!bot, Mbell, Headbomb, Escarbot, Maliz, Fastman99, Felixbecker2, Cuzkatzimhut, Red Act, Lejarrag, StevenJohnston, YohanN7, Bobathon71, EverettYou, SchreiberBike, Gedankenpause, Addbot, Luckas-bot, Yobot, MichalKotowski, Tamtamar, WWsemir, Xqbot, Pra1998, Llsalcedo, Topherwhelan, EmausBot, Stibu, Maschen, ChuispastonBot, Helpful Pixie Bot, BG19bot, Ema--or, Enyokoyama, Lý Minh Nhật, LeiFang86 and Anonymous: 48

- **Interaction picture***Source:*https://en.wikipedia.org/wiki/Interaction_picture?oldid=696724174*Contributors:*Charles Matthews, Phys, Giftlite, Elroch, Sam Hocevar, Pjacobi, Laurascudder, Jess Riedel, Sbyrnes321, SmackBot, Colonies Chris, Fuhghettaboutit, Colin Kiegel, Pinval~enwiki, Shiznick, Sheliak, Cuzkatzimhut, Thurth, Pamputt, Alexbot, Addbot, LaaknorBot, Luckas-bot, Yobot, Kaveh.khodjasteh, Citation bot, 🔲🔲, Kismalac, Kirsim, Trappist the monk, Butsurigaku, Quondum, Anagogist, BG19bot, Magicsowon, Vieque and Anonymous: 19

- **Schrödinger picture** *Source:* https://en.wikipedia.org/wiki/Schr%C3%B6dinger_picture?oldid=693591084 *Contributors:* Michael Hardy, Cyp, AugPi, Tobias Bergemann, Ancheta Wis, Lethe, Masudr, Pjacobi, Plugwash, Laurascudder, Jérôme, Gene Nygaard, Linas, StradivariusTV, Ryan Reich, NawlinWiki, Jess Riedel, RG2, SmackBot, Bluebot, Colonies Chris, Javalenok, Pen of bushido, Jaapkroe, Loadmaster, CmdrObot, Mct mht, Magioladitis, BigrTex, Soumya m, Sheliak, Cuzkatzimhut, VolkovBot, Thurth, Pamputt, Kbrose, YohanN7, Hugh16, Likebox, Gorchy, MystBot, Addbot, Looie496, Luckas-bot, Yobot, PowerUserPCDude, 🔲🔲, Kismalac, Tom.Reding, Seattle Jörg, Jordgette, Solomonfromfinland, ZéroBot, Cogiati, Maschen, Zueignung, Iuliuscurt, YFdyh-bot, JYBot, Cesaranieto~enwiki, Mark viking, Airwoz and Anonymous: 30

- **Heisenberg picture** *Source:* https://en.wikipedia.org/wiki/Heisenberg_picture?oldid=695116012 *Contributors:* Bdesham, Michael Hardy, Docu, Charles Matthews, Patrick0Moran, Phys, Bkalafut, Tobias Bergemann, Giftlite, Lethe, MathKnight, Fastfission, Lumidek, Pj.de.bruin, Pjacobi, Laurascudder, Eruantalon, DV8 2XL, Oleg Alexandrov, Hfarmer, -Ril-, GeorgeOrr, GregorB, SeventyThree, MarSch, Mathbot, Privong, Chobot, GangofOne, Naar, No3, Larsobrien, Jess Riedel, SmackBot, ElTchanggo, Colonies Chris, Javalenok, QFT, Georg-Johann, Zarniwoot, Zipz0p, Karl-H, Headbomb, Karibas, Limakom, Magioladitis, Cobaltnova, Soumya m, Sheliak, Cuzkatzimhut, Thurth, Joerglwitsch, Pamputt, Klappspatier, Neparis, Kbrose, YohanN7, Jam343, Likebox, ClueBot, 1ForTheMoney, Dark Mage, MystBot, Addbot, Мыша, Looie496, Luckas-bot, Yobot, Sirsparksalot, Citation bot, PowerUserPCDude, 🔲🔲, Kismalac, Tank00, Mtrencseni, Brazmyth, Quondum, Ohyoungloo, Zueignung, Helpful Pixie Bot, Frze, Mogism, Mark viking, YiFeiBot, Monkbot, Mattster1995, Erjas3oa and Anonymous: 49

10.8.2 Images

- **File:AM_diagrams_bra.svg** *Source:* https://upload.wikimedia.org/wikipedia/commons/3/3e/AM_diagrams_bra.svg *License:* CC0 *Contributors:* Own work *Original artist:* Maschen

- **File:AM_diagrams_bra_time_reversed.svg***Source:*https://upload.wikimedia.org/wikipedia/commons/2/22/AM_diagrams_bra_time_rev svg *License:* CC0 *Contributors:* Own work *Original artist:* Maschen

- **File:AM_diagrams_inner_product.svg** *Source:* https://upload.wikimedia.org/wikipedia/commons/5/51/AM_diagrams_inner_product.svg *License:* CC0 *Contributors:* Own work *Original artist:* Maschen

- **File:AM_diagrams_inner_product_contraction.svg** *Source:* https://upload.wikimedia.org/wikipedia/commons/7/74/AM_diagrams_inner_product_contraction.svg *License:* CC0 *Contributors:* Own work *Original artist:* Maschen

- **File:AM_diagrams_inner_product_of_tensor_product.svg** *Source:* https://upload.wikimedia.org/wikipedia/commons/f/f4/AM_diagrams_inner_product_of_tensor_product.svg *License:* CC0 *Contributors:* Own work *Original artist:* Maschen

- **File:AM_diagrams_inner_product_of_tensor_product_time_reversed.svg** *Source:* https://upload.wikimedia.org/wikipedia/commons/2/2a/AM_diagrams_inner_product_of_tensor_product_time_reversed.svg *License:* CC0 *Contributors:* Own work *Original artist:* Maschen

- **File:AM_diagrams_inner_product_time_reversed.svg***Source:*https://upload.wikimedia.org/wikipedia/commons/b/b3/AM_diagrams_in product_time_reversed.svg *License:* CC0 *Contributors:* Own work *Original artist:* Maschen

- **File:AM_diagrams_ket.svg** *Source:* https://upload.wikimedia.org/wikipedia/commons/d/df/AM_diagrams_ket.svg *License:* CC0 *Contributors:* Own work *Original artist:* Maschen

- **File:AM_diagrams_ket_time_reversed.svg***Source:*https://upload.wikimedia.org/wikipedia/commons/0/05/AM_diagrams_ket_time_rever svg *License:* CC0 *Contributors:* Own work *Original artist:* Maschen

- **File:AM_diagrams_outer_product.svg** *Source:* https://upload.wikimedia.org/wikipedia/commons/2/26/AM_diagrams_outer_product.svg *License:* CC0 *Contributors:* Own work *Original artist:* Maschen

- **File:AM_diagrams_outer_product_contraction.svg** *Source:* https://upload.wikimedia.org/wikipedia/commons/3/31/AM_diagrams_outer_product_contraction.svg *License:* CC0 *Contributors:* Own work *Original artist:* Maschen

- **File:AM_diagrams_outer_product_time_reversed.svg***Source:*https://upload.wikimedia.org/wikipedia/commons/b/b0/AM_diagrams_ou product_time_reversed.svg *License:* CC0 *Contributors:* Own work *Original artist:* Maschen

- **File:AM_diagrams_tensor_product_kets.svg** *Source:* https://upload.wikimedia.org/wikipedia/commons/5/59/AM_diagrams_tensor_product_kets.svg *License:* CC0 *Contributors:* Own work *Original artist:* Maschen

- **File:AM_diagrams_tensor_product_kets_time_reversed.svg***Source:*https://upload.wikimedia.org/wikipedia/commons/f/fb/AM_diagra tensor_product_kets_time_reversed.svg *License:* CC0 *Contributors:* Own work *Original artist:* Maschen

- **File:Beta_Negative_Decay.svg** *Source:* https://upload.wikimedia.org/wikipedia/commons/8/89/Beta_Negative_Decay.svg *License:* Public domain *Contributors:* This vector image was created with Inkscape. *Original artist:* Joel Holdsworth (Joelholdsworth)

- **File:BosonFusion-Higgs.svg** *Source:* https://upload.wikimedia.org/wikipedia/commons/7/78/BosonFusion-Higgs.svg *License:* CC-BY-SA-3.0 *Contributors:*

- BosonFusion-Higgs.png *Original artist:* BosonFusion-Higgs.png: User:Harp 12:43, 28 March 2007

10.8.3   Content license

www.ingramcontent.com/pod-product-compliance
Lightning Source LLC
Chambersburg PA
CBHW080708190526
45169CB00006B/2294